Stroke Rehabilitation in Low-Resource Settings

Dorcas B. C. Gandhi • Jeyaraj D Pandian
Editors

Stroke Rehabilitation in Low-Resource Settings

Evidence Implementation

Springer

Editors
Dorcas B. C. Gandhi
Department of Neurology, Christian Medical College & Hospital Ludhiana, Ludhiana, Punjab, India

Department of Physiotherapy, Manipal Academy of Higher Education, Manipal, Karnataka, India

Jeyaraj D Pandian
Department of Neurology
Christian Medical College and Hospital
Ludhiana, Punjab, India

ISBN 978-981-92-0402-1 ISBN 978-981-92-0403-8 (eBook)
https://doi.org/10.1007/978-981-92-0403-8

This Springer imprint is published by the registered company Springer Nature Singapore Pte Ltd.
The registered company address is: 152 Beach Road, #21-01/04 Gateway East, Singapore 189721, Singapore

Dedicated to my mother, Mrs. Pushpa Gandhi, whose example taught me the value of discipline and perseverance, and the joy of learning and striving for excellence

Foreword

In my early 20's I travelled to India for my last clinical rotation of my Australian physiotherapy degree. It was 1984, the first year that CMC Vellore accepted Australian students. A formative experience, both personally and professionally, I learned much in my 3-month stay. One major takeaway from the experience was the ingenuity of staff in this country of high population and low health resources to create simple, scalable solutions to equipment needs where shortages abound.

Forty years on, it is a privilege to write this foreword to a book that showcases the progress that has been made in stroke rehabilitation research and practice, not just in low- to middle-income countries (LMICs), but worldwide.

I congratulate the Editors, Dorcas Gandhi and Jeyaraj Pandian, and the authorship team on the scope and format of the book. Over 12 chapters, they explore a range of stroke rehabilitation service models that have been applied to fill current gaps in LMICs stroke care, providing examples of models that have successfully met rehabilitation needs in a range of settings and across multiple countries. Core areas of therapy, training or support to address common post-stroke impairments, including speech and language deficits and pain management are presented. Advice on accessible, evidence-based training packages for specific impairments (for example, GRASP for upper limb training, Chap. 3) or self-management programs (for example, Take Charge, Chaps. 10 and 11), to name just a few, are offered throughout. In each chapter the team have focused on presenting evidence to support practice, efforts to contextualise the evidence to LMICs, and provide simple summaries of key takeaways. Importantly, these summaries contain not only what to do, but what to stop doing. These messages are critical. In all health services, anywhere in the world, we should be mindful that the time we spend with patients is used to deliver the best-evidence, most effective care possible. Time is valuable, both that of our patients and ourselves. In countries where trained health professionals and equipment are often limited, the time that health professionals spend with patients and families, or with other care providers trained to support rehabilitation, is even more important. The key message here is to spend time on the things most likely to make a difference to the daily lives of our patients. This book will help readers to do just that.

The importance of the multidisciplinary team, and respect for all members, is a recurring theme in the book, starting in Chap. 1. The world-wide shortage of health professional's expert in stroke rehabilitation is a major challenge in LMICs. Flexibility, rather than dogged protection of professional boundaries, and creating systems to boost the rehabilitation workforce under these conditions, are vital to meet the challenges of delivering stroke rehabilitation in LMICs. The need to make rehabilitation visible and for Government and professional bodies to support this mammoth task through licensing of health professionals and adequate funding of rehabilitation services is clearly made. 'Task-shifting', the pragmatic idea that family and other unskilled workers can provide rehabilitation when skilled staff are in short supply, is raised in several chapters. Examples are provided where task-shifting models have been successfully implemented. The authors are careful to note the limits of current evidence to support these models, while also offering ideas for supervised care models that maintain a degree of expert input.

The need for home or community-based rehabilitation services to support recovery over the longer term is well recognised in this book. The authors rightly question whether reproducing established, expensive, inpatient care models, common in HIC, is the right approach for LMICs with no or limited stroke rehabilitation. The potential for technology to support recovery after stroke through a range of solutions ranging from simple telehealth, to lower-cost mHealth or robotic options, is therefore a strong theme throughout the book. As with all technological solutions, in both high and LMICS, demonstrating that these solutions are feasible in a range of countries, at scale, sustainable beyond the end of a research cycle, and lead to meaningful improvements in outcomes is the big challenge.

This group of authors represent a new generation of researchers and clinicians who are leading by example to demonstrate that through collaboration, information and resource sharing, and evidence contextualisation for LMICs, it is possible to accelerate change. I am excited by their thoughtful and ambitious vision for improving stroke rehabilitation in LMICs and their potential to lead that change and bring others along with them.

If you are just beginning your stroke rehabilitation practice, this book provides an excellent starting point, outlining common barriers, but also how to overcome them and what to implement in your own environment. If you are an experienced health professional, it offers you an up-to-date guide of best practice. I invite you to read this book and become part of this exciting collaborative movement for change.

Much has changed over my 40-year career in stroke rehabilitation and recovery. Stroke remains an exciting, yet challenging, area of work and study. While we have many proven interventions in our rehabilitation tool kit, we can still do better, and more research is needed. Most importantly, if we could effectively implement what we already know to work to all those in need, we could lower the high burden of stroke-related disability, which sits most heavily in LMICs. My hope is that this book helps readers feel empowered to take action in their own practice or sphere of influence to improve stroke rehabilitation.

Julie Bernhardt, AM, FAHMS, PhD

Stroke Theme
Florey Institute of Neuroscience
and Mental Health
Melbourne, VIC, Australia

Preface

After stroke, survival is only the beginning; rehabilitation is what turns survival into dignity, function, and hope

Stroke is one of the leading causes of long-term disability worldwide, and its burden is increasingly concentrated in low- and middle-income countries (LMICs), where health systems often face significant constraints in infrastructure, workforce, financing, and continuity of care. In such settings, stroke rehabilitation is not merely a clinical service; it is a public health necessity, a social support imperative, and, very often, a test of how equitably a health system functions. This book was conceived in response to that reality. Stroke rehabilitation is a right for all stroke survivors and should not be dependent of where the survivor lives or what therapy he or she can afford.

The motivation for this volume lies not only in epidemiology or policy, but in everyday clinical experience across LMICs. Many stroke survivors in these settings survive the acute event but face a second crisis soon after discharge: limited access to rehabilitation, inadequate follow-up, financial hardship, shortages of trained professionals, and caregiving demands that fall heavily on families. The result is that disability after stroke is often amplified not only by the neurological injury itself, but by the context in which recovery must unfold. Several chapters in this book reflect this reality directly, highlighting the challenges of service access, discharge transitions, community reintegration, and the need for practical models of care suited to low-resource environments.

A familiar story from many LMICs helps explain why this book became necessary. A middle-aged stroke survivor from a semi-urban or rural community may receive life-saving acute treatment, yet return home with persistent weakness, communication difficulty, swallowing problems, fatigue, cognitive changes, and reduced independence in daily activities. The nearest rehabilitation centre may be distant, transport costs high, specialist services scarce, and structured home-based follow-up unavailable. Recovery then depends not only on formal rehabilitation, but on family-led care, improvised adaptations, community support, and the ability of clinicians to translate evidence-based principles into feasible local practice. That gap between what is ideal and what is achievable is where this book is situated.

This volume does not assume that rehabilitation models developed in high-resource settings can simply be transplanted into LMICs without adaptation.

Instead, it recognizes that contextualisation is central to effective stroke rehabilitation. Across its chapters, the book addresses the acute phase of rehabilitation, upper limb and gait recovery, cognitive and perceptual rehabilitation, speech-language and swallowing rehabilitation, activities of daily living, often neglected post-stroke complications, technology-enabled but affordable interventions, discharge planning, community and low-cost service models, and practical standard operating procedures for hospital and community settings. The unifying principle across these contributions is clear: meaningful rehabilitation in LMICs must be evidence-informed, function-focused, culturally responsive, and feasible within local constraints.

Importantly, this book is not written from a position of resignation. It is grounded in the conviction that high-quality stroke rehabilitation does not always require high-cost solutions. In many low-resource settings, progress is made through coordinated multidisciplinary work, task-sharing, family engagement, community-based rehabilitation, early supported discharge where feasible, and the careful use of low-cost technologies and locally available materials. Such approaches are not substitutes for system strengthening; they are pragmatic pathways toward it.

This book is intended for clinicians, rehabilitation professionals, educators, researchers, programme planners, and policymakers who are engaged in improving stroke care in resource-constrained settings. It aims to support both practice and implementation: not only what should be done, but what can be done, and how. If it succeeds, it will help bridge the persistent gap between evidence and access, and between survival and meaningful recovery.

Above all, this book is shaped by the experiences of stroke survivors and their families across LMICs. Their struggles, adaptations, and resilience have been the strongest argument for bringing this work together. This volume, with its appendix, is offered in the hope that it will contribute, in a practical and lasting way, to more accessible, more context-responsive, and more humane stroke rehabilitation.

Ludhiana, Punjab, India Dorcas B. C. Gandhi

Co-editor Preface

My work in stroke care has taken me across hospitals and communities in Africa, South and Southeast Asia, Latin America, and other parts of the world. In these settings, I have met stroke survivors discharged home within days of their event, families suddenly tasked with becoming therapists, and clinicians striving to deliver evidence-based care with limited time, personnel, and equipment. I have also witnessed remarkable ingenuity—rehabilitation delivered in courtyards, walking practice along village roads, communication therapy improvised through song, gesture, and family storytelling. These encounters have profoundly shaped my understanding of what stroke rehabilitation must look like in low-resource settings: grounded in evidence, yes, but deeply contextual, pragmatic, and people-centred.

Globally, stroke remains a leading cause of long-term disability, with the vast majority of affected individuals living in low- and middle-income countries (LMICs). In these regions, the burden of disability is compounded by weak rehabilitation systems, limited workforce capacity, fragmented transitions of care, and minimal access to long-term support. Rehabilitation is often the least developed component of stroke services, despite its central role in recovery, participation, and quality of life. This inequity stands in direct contrast to global commitments that recognize rehabilitation as an essential health service.

As President of the World Stroke Organization (WSO), I am guided by the WSO Global Stroke Roadmap, WSO stroke guidelines and the Lancet Neurology WSO Commission pragmatic recommendations to reduce the global burden of stroke documents, which emphasize stroke rehabilitation as a core component of quality stroke care across the entire continuum—from the acute phase through community reintegration. The WSO framework explicitly acknowledges that while not all regions can deliver comprehensive services immediately, all settings can implement tiered, context-appropriate rehabilitation interventions that improve outcomes and reduce disability. This book directly operationalizes that vision by translating evidence into feasible action for LMIC contexts, aligned with real-world constraints and opportunities.

This volume is also firmly anchored in the WHO Rehabilitation 2030: A Call for Action, which positions rehabilitation as indispensable to universal health coverage and sustainable development. Rehabilitation 2030 highlights the profound unmet need for rehabilitation globally—particularly in LMICs—and calls for integration of rehabilitation into all levels of health systems, development of context-relevant service delivery models, strengthening of the rehabilitation workforce, and improved continuity of care across hospital and community settings. Stroke rehabilitation exemplifies both the urgency and the opportunity embedded in this agenda.

The chapters in this book reflect these global priorities through a distinctly Global South lens. They address post-stroke rehabilitation in the acute phase, not as a specialist luxury, but as a set of essential actions that can be embedded even within basic inpatient care. Contextualised approaches to arm and hand function, gait and balance, cognition, perception, communication, swallowing, and activities of daily living emphasize task-oriented, family-enabled, and resource-conscious strategies. These chapters recognize the central role of caregivers, community health workers, and primary care systems in sustaining recovery long after hospital discharge.

Equally important is the attention given to unaddressed post-stroke complications—pain, fatigue, depression, contractures, and social isolation—which disproportionately affect stroke survivors in LMICs and remain largely invisible in overstretched systems. The focus on discharge planning and transition of care speaks directly to one of the most critical failures in stroke services in low-resource settings: the absence of structured pathways linking hospital-based care to community and home-based rehabilitation.

Technology-enhanced rehabilitation is explored through a lens of equity and feasibility, aligned with both WSO and WHO priorities. Rather than promoting high-cost solutions, this book examines how mobile technologies, tele-rehabilitation, and digital education platforms can extend reach, support supervision, and improve continuity of care without deepening existing disparities. The discussion on models of stroke rehabilitation in low-resource settings reflects the ground realities of LMIC health systems and underscores the need for adaptable, scalable, and locally owned solutions.

This book is ultimately a call to action. It calls on clinicians to adapt evidence with confidence, on educators to train a workforce fit for context, on health systems to integrate rehabilitation across levels of care, and on policymakers to recognize rehabilitation as an investment in human capital and social development—precisely as articulated in the WHO Rehabilitation 2030 agenda.

For the millions of stroke survivors and caregivers across the LMICs, recovery is not merely a clinical outcome—it is the ability to walk back into their communities, communicate with dignity, return to meaningful roles, and reclaim their place in society. It is my hope that this book will contribute, in a practical and tangible way, to making that recovery possible—everywhere.

Ludhiana, Punjab, India Jeyaraj Durai Pandian

Contents

1 Stroke Rehabilitation Burden and Unique Features in Low-Resource Settings

Vinicius Montanaro, Dorcas B. C. Gandhi, and Rinita Mascarenhas

Abstract

The chapter reviews core evidence-based practices in stroke rehabilitation and discusses strategies for adapting these interventions to environments with workforce shortages, financial constraints, and limited infrastructure. It examines the role of multidisciplinary teams and presents viable alternatives, including task shifting, caregiver involvement, and community-based models of care, to ensure continuity across the recovery pathway. Emerging innovations such as telerehabilitation, recent technologies, robotics, and virtual reality are also discussed, with an emphasis on their feasibility, scalability, and contextual adaptation for low-resource settings. The chapter further delves into actionable implementation frameworks and policy considerations to guide clinicians, administrators, and policymakers in strengthening stroke rehabilitation systems. Integrating evidence-based practice with context-sensitive innovation enables sustainable improvements in stroke recovery across diverse healthcare settings.

V. Montanaro (✉)
Neurological Rehabilitation SARAH Network of Rehabilitation Hospitals, Brasília, Brazil

Radbou University Medical Center, Nijmegen, Netherlands

Programa de Pós Grduação em Neurologia da Universidade Federal Fluminense, Niterói, Brazil

SARAH Network of Rehabilitation Hospitals, Brasília, Brazil

D. B. C. Gandhi
Department of Neurology, Christian Medical College & Hospital Ludhiana, Ludhiana, Punjab, India

Department of Physiotherapy, Manipal Academy of Higher Education, Manipal, Karnataka, India

R. Mascarenhas
Department of Neurology, Christian Medical College and Hospital, Ludhiana, Punjab, India

D. B. C. Gandhi, J. D Pandian (eds.), *Stroke Rehabilitation in Low-Resource Settings*, https://doi.org/10.1007/978-981-92-0403-8_1

Keywords

Low- and middle-income countries · Evidence-based practice · Multidisciplinary rehabilitation · Stroke rehabilitation models · Scalabilty

1.1 Introduction

Stroke remains one of the leading causes of disability worldwide, and its impact is disproportionately felt in low- and middle-income countries (LMICs). According to the Global Burden of Disease (GBD) study, over 70% of strokes and 80% of stroke-related deaths occur in LMICs, where resources for rehabilitation are often scarce and inaccessible. The years lived with disability (YLDs) attributed to stroke in LMICs have risen significantly over the past decade, reflecting both increasing stroke incidence and the lack of adequate post-stroke care systems [1].

Rehabilitation in these settings is hampered by a myriad of challenges, including insufficient healthcare infrastructure, a dearth of trained professionals, and limited financial and technological resources. Furthermore, cultural and systemic barriers often impede the timely initiation of rehabilitation, exacerbating long-term disability and economic dependency. This chapter explores the unique burden of stroke rehabilitation in LMICs, emphasizing evidence-based strategies, innovative solutions, and contextual adaptations to address these challenges.

1.2 Evidence-Based Strategies for Optimizing Outcomes

Stroke rehabilitation in low-resource settings presents a distinct set of challenges and opportunities for optimizing patient outcomes. In these settings, limited access to specialized care, rehabilitative technologies, and multidisciplinary teams often complicates recovery. Despite these challenges, evidence-based practices (EBPs) tailored to resource-constrained environments offer viable pathways to improve stroke recovery outcomes. This section provides an overview of EBPs in stroke rehabilitation, explores their adaptation to low-resource settings, and highlights successful case studies, underscoring the potential of strategic and innovative interventions to improve patients' quality of life.

1.2.1 Overview of Evidence-Based Practices in Stroke Rehabilitation

Evidence-based practices (EBPs) in stroke rehabilitation prioritize targeted interventions that have demonstrated efficacy through rigorous research. Core strategies include task-specific training, constraint-induced movement therapy (CIMT), motor learning principles, and compensatory strategies that focus on adaptive functions for daily living. Research indicates that task-specific and repetitive practice significantly improves motor function, balance, and cognitive recovery in post-stroke patients. Key guidelines from leading organizations, such as the American Heart Association and the World Stroke Organization, emphasize an interdisciplinary approach, combining physiotherapy, occupational therapy, and speech therapy to address the multifaceted effects of stroke [1, 2].

Neuroplasticity-based interventions, which leverage the brain's capacity to reorganize after injury, remain central to EBPs. Repetitive, task-oriented training is particularly beneficial in promoting neural recovery, especially within the initial 6 months post-stroke when neuroplastic potential is highest. With adequate implementation, these EBPs improve physical function, reduce disability, and enhance the likelihood of reintegration into daily life, providing foundational strategies for rehabilitation practitioners in various settings [3].

1.2.2 Adaptation to Low-Resource Settings

Implementing EBPs in low-resource environments requires careful adaptation to maximize available resources and local capacity. In these settings, constraints such as limited equipment, professional staff shortages, and geographical barriers to care necessitate creative solutions. For instance, tele-rehabilitation, facilitated by basic mobile phone technology, can bridge geographical gaps and enable remote monitoring and guidance for patients unable to travel. Studies in rural India and sub-Saharan Africa have shown that tele-rehabilitation is an effective alternative, allowing patients to benefit from expert supervision and adaptive exercise programs, even without traditional rehabilitative resources [4].

Community-based rehabilitation (CBR) is another adaptive model that empowers local health workers and family members to deliver care under the guidance of remote specialists. This model is increasingly recognized for its efficacy in low-resource settings, where traditional rehabilitation facilities may be scarce. Additionally, "task-shifting" initiatives, where basic rehabilitative tasks are delegated to community health workers, have been effective in regions with limited specialist availability. For example, in Kenya and rural parts of South America, trained community health workers provide structured exercise routines and cognitive therapies, enabling early intervention that is otherwise inaccessible [5].

Furthermore, using inexpensive, locally sourced materials to create rehabilitative aids, such as makeshift exercise bands or modified utensils, has been shown to support functional recovery. A study conducted in Mozambique adapted CIMT techniques using minimal equipment, resulting in improved upper-limb function among participants without incurring prohibitive costs [6]. These adaptations underscore the value of flexible, low-cost interventions that align with EBP frameworks yet remain accessible to underserved populations.

1.2.3 Case Studies Demonstrating Successful Outcomes

Several case studies exemplify how tailored EBPs, even with limited resources, can lead to meaningful improvements in stroke rehabilitation outcomes.

1. Tele-rehabilitation in rural India
   In a case study in Tamil Nadu, India, stroke patients participated in a 6-week tele-rehabilitation program using a low-bandwidth mobile application. Through this app, physical therapists guided patients in functional exercises and strength-building routines. The program significantly improved patients' motor and functional abilities, demonstrating the potential of tele-rehabilitation to circumvent geographical and infrastructural limitations. By delivering effective rehabilitation remotely, the study highlighted a cost-effective model that can be scaled across similar regions [4].
2. Community-based rehabilitation in Malawi
   In Malawi, a community-based rehabilitation program supported by local health workers was implemented for post-stroke patients in a rural district. This initiative trained community health workers to provide home-based rehabilitation, focusing on mobility and self-care tasks. Over 6 months, patients showed marked improvements in mobility and activities of daily living (ADL), along with substantial reductions in caregiver burden. This success underlines the capacity of community health workers to deliver EBPs, achieving significant rehabilitative outcomes despite limited professional resources [7].
3. Constraint-induced movement therapy (CIMT) in Mozambique
   An innovative adaptation of CIMT in Mozambique involved using locally sourced materials to support upper-limb rehabilitation. Participants with hemiparesis were encouraged to use their affected limbs in daily tasks, while their unaffected limbs were restricted. Despite limited clinical oversight and resources, participants experienced a 40% increase in upper-limb functionality over 3 months. This model demonstrates that even resource-intensive EBPs like CIMT can be successfully adapted to low-resource environments with creative modifications [6, 8].
4. WHO training in Bhutan: stroke care and rehabilitation
   The disability-adjusted life years (DALYs) due to stroke in Bhutan are lower than those in most other countries in South Asia. They implemented standardized stroke care protocols, completed training in establishing stroke registries, and strengthened coordination among healthcare teams for timely stroke intervention through a collaboration between the World Health Organization (WHO) and Christian Medical College, Ludhiana (CMCL). Additionally, standardized protocols for patient rehabilitation were established and enhanced to improve overall care outcomes. This, in turn, led to strengthened coordination among healthcare teams for timely stroke intervention.

   Established and improved standardized protocols for patient rehabilitation.

We can conclude that evidence-based stroke rehabilitation practices, adapted to the realities of low-resource settings, offer hope for improving patient outcomes and quality of life. These adaptations—ranging from tele-rehabilitation to community-based models and resourceful modifications of standard therapies—highlight the resilience and adaptability required in these settings. While further research and support are essential to optimize these models, existing evidence underscores that well-adapted EBPs can bridge the gap in care and deliver tangible, life-changing results for stroke patients in low-resource settings.

1.3 Multidisciplinary Rehabilitation Team Dynamics

A multidisciplinary team (MDT) is simply the coordinated efforts of various specialized healthcare professionals with the overarching goal of improving treatment efficiency and quality of care provided to patients [9]. This approach to service delivery first emerged in the 1980s in the field of oncology, but soon spread across other care domains due to its proven effectiveness in improving survival rates. In the world of stroke, MDTs include rehabilitation specialists; rehabilitation nurses; physical, occupational, recreation, speech, and language therapists; mental health professionals; social workers; and patients and their caregivers. Each member of the team utilizes the skills and expertise of their discipline in managing the patient from their perspective. This may result in separate individual consultations with these specialists who meet regularly in the absence of the patient to discuss findings and future directions for the patient. Such teams certainly provide more evidence-based knowledge and experience than when the disciplines work in isolation. While various disciplines bring their expertise in a MDT approach integrating separate disciplines, an interdisciplinary team (IDT) can integrate such distinct expertise to provide a collaborative goal setting, care planning, and decision-making in which the patient and caregivers are an active part [10]. A holistic approach to cater to all aspects of the patient's care is key to the IDT approach, which in turn empowers the patient by being part of the decision-making process—a patient-centered approach. Additionally, each discipline is encouraged to question one another and explore the best and alternative solutions for the best outcome for the patient [11].

1.3.1 Viable Alternatives to MDT or IDT Approaches in Low-Resource Settings

An IDT approach is one of the key quality indicators of the acute phase (1–7 days post-stroke onset) in an essential and advanced stroke unit as described by the World Stroke Organization. They describe an IDT as a team with at least a physician with training in stroke, a nurse, and a rehabilitation specialist (like physiotherapist, occupational therapist, or speech-language therapist). In a low-resource setting, this minimum standard of team composition can be relied on in the absence of specialists from all disciplines, as mentioned previously [12].

Such a task-sharing or task-shifting approach among rehabilitation professionals is being widely used across LMICs even when it is not reported frequently. One such example is from an upper-middle-income country, China, where the application of a nurse-led rehabilitation program for those with stroke and limb dysfunction showed significant improvement in upper and lower limb scores on the Fugl–Meyer Assessment scale and in quality of life during their in-hospital stay with no adverse events [13]. In many hospitals, stroke team coordinators function as a make-do MDT, acting as a bridge between the patient/caregiver and the wider MDT from admission until discharge and, in some cases, following up with the patient in the community at regular intervals.

1.3.2 Importance of MDT/IDT, Teamwork, and Communication for Seamless Care Provision and Transition

The benefits of an MDT/IDT in stroke care are undeniable, with improved outcomes, reduced death or dependency, and decreased need for institutional care. Evidence suggests that such an approach assists in earlier regaining functional independence in activities of daily living, reduces hospital length of stay, and post-stroke morbidity [14]. An IDT allows team members to work as equals, valuing the knowledge and expertise brought by each discipline. IDT members understand each other's roles and accept overlap and blurred boundaries in certain situations, which allows timely information exchange, effective and speedy implementation of rehabilitative care, and seamless transitions to the community.

Communication and teamwork are key to smooth functioning of an IDT. This can be facilitated by maintaining an environment of equality among professionals and establishing easy communication channels. There are various ways in which a communication channel can be structured. The use of predefined template/performa for recording patient reports and progress that are accessible to all team members is one such approach. Alternately, mobile-based applications, tablet-based technology, or computer-based systems can be implemented to enable all team members to access patient-specific details. In low-resource settings, weekly meetings of the IDT in the presence of the patient/caregiver or a simple WhatsApp-based discussion among team members can be initiated.

Successful teamwork can be established through open communication, a nonpunitive environment, clear direction, clear and known roles and tasks for team members, a respectful atmosphere, shared responsibility for team success, appropriate balance of member participation for the task at hand, acknowledgment and processing of conflict, clear specifications regarding authority and accountability, clear and known decision-making procedures, regular and routine communication and information sharing, and an enabling environment, including access to needed resources, mechanism to evaluate outcomes, and adjust accordingly [14–16].

Challenges associated with MDT/IDT also need to be addressed, especially when they are in the initial phase of implementation. Team members need to understand and learn to work with conflicting priorities/goals, limited resources that need to be shared, taking over the responsibilities of certain disciplines when they are not available in a low-resource setting, and time constraints [14, 15].

1.3.3 MDT/IDT in the Community Settings

Beyond the in-hospital phase post-stroke, a comprehensive and multidisciplinary care is essential in the community settings for patients. Such a service is usually accessed via outpatient departments of hospitals, clinics, or rehabilitation centers. In low-resource settings, however, congregation of such centers may be limited to urban areas or unavailable at all, with disciplines working in silos. In a recent reviewby An Z et al., the MDT approach in communities was delivered by a combination of hospital-based and community-based teams, with physicians in hospitals and general practitioners in communities having the responsibility of overseeing the program, providing guidance, and evaluating outcomes [17]. With the assistance of a team that majorly included rehabilitation therapists, nurses, and psychologists, community-based programs focused on health promotion, prevention, medical care, rehabilitation, psychological guidance, and assistive devices. However, the studies in this review were conducted in the United Kingdom, South Korea, France, and Canada. In low-resource settings, such a coordinated effort seems difficult at this stage, with a huge load of patient per neurologist/physician, leading to time constraints in overlooking such community-based programs. Alternatively, nurse-led programs have been explored, with sufficient training in rehabilitation and its outcomes, with the support of social workers in the community. The remote delivery of rehabilitation programs can leverage self-guided learning materials, periodic check-ins via phone calls, and synchronous tele-rehabilitation sessions through platforms like WhatsApp or Zoom. These alternative methods of rehabilitation, which can be overseen by community health workers, social workers, or even trained caregivers, offer accessible and cost-effective solutions for low-resource settings. Such programs have proven useful in Malawi, Brazil, Benin, and India.

1.4 Recent Innovations in Stroke Rehabilitation

In recent years, stroke rehabilitation has benefited from a surge of innovations designed to optimize recovery, improve quality of life, and, importantly, adapt to the unique demands of low- and middle-income countries (LMICs) [18]. As the burden of stroke grows in LMICs, limited access to specialist care and rehabilitative technologies has prompted researchers and practitioners to explore adaptable, cost-effective solutions. This section provides an overview of cutting-edge approaches in stroke rehabilitation, highlights key findings from recent research, and contextualizes these innovations for low-resource environments.

1.4.1 Overview of Novel Approaches and Technologies in Stroke Rehabilitation

Emerging technologies in stroke rehabilitation have broadened the scope of traditional practices, incorporating techniques and tools designed to facilitate neural recovery, improve functional outcomes, and engage patients in meaningful, self-directed exercises. These innovations include:

1. Tele-rehabilitation and mobile health (mHealth): Tele-rehabilitation and mHealth are transforming the accessibility of stroke rehabilitation, particularly in rural or underserved areas. Through digital platforms, patients can engage in tailored exercises, receive feedback, and participate in virtual consultations with specialists. Tele-rehabilitation interventions provide consistent access to care, with studies indicating outcomes comparable to in-person therapy [19]. mHealth applications also enable self-management and remote monitoring, fostering patient autonomy and continuity of care across geographical barriers.
2. Wearable technology and biometric feedback: Advances in wearable devices now enable real-time monitoring of key parameters, including gait, muscle activity, and balance, which are vital for post-stroke recovery. Devices like smart insoles, motion sensors, and wrist-based accelerometers collect movement data, providing immediate feedback on progress and identifying areas for improvement. These devices have shown promising results in refining motor skills and reinforcing balance training, particularly in self-guided home settings, which is critical in LMICs where professional supervision may be limited [20].
3. Robotics and exoskeletons: Robotic devices and exoskeletons are now playing a critical role in facilitating repetitive and precise motor exercises, which are central to neuroplastic recovery. Exoskeletons, often attached to arms or legs, support limb movement by enhancing muscle activation patterns, which encourages neuroplasticity. While high costs and complex maintenance have limited their widespread use, the development of more affordable, simplified versions is gradually making robotic-assisted rehabilitation more accessible to LMICs [21].
4. Virtual reality (VR) and augmented reality (AR): VR and AR offer immersive, interactive environments where stroke survivors can engage in simulated activities, enhancing their motivation and providing real-time feedback. These technologies support motor learning by replicating everyday tasks and reinforcing functional skills in a safe, controlled setting. Research demonstrates that VR-based training significantly improves upper-limb function and balance, particularly when conventional methods are supplemented with VR [22]. Portable, low-cost VR systems are now being explored to make this technology accessible in lower-resource regions.

1.4.2 Analysis of Recent Research Findings

Recent studies underscore the efficacy of these technologies in enhancing recovery outcomes, particularly when integrated into traditional rehabilitation frameworks.

- Tele-rehabilitation: A meta-analysis of tele-rehabilitation programs across LMICs revealed substantial improvements in patient functional outcomes, quality of life, and adherence rates [23]. This research highlights that tele-rehabilitation achieves efficacy comparable to in-person treatment, particularly in motor skill and daily functional task recovery. In a recent randomized controlled trial conducted in South Africa, stroke survivors using a tele-rehabilitation platform displayed significant improvements in motor function and self-care compared to a control group receiving standard in-person therapy once per month [24]. Such findings emphasize the value of technology-mediated rehabilitation in overcoming geographic and economic barriers. However, there is a need for additional training that is structured and introduced into graduate rehabilitation programs for the smooth delivery of tele-rehabilitation [25, 26].
- Wearable technology: Clinical trials of wearable devices indicate marked improvements in patients' gait, balance, and muscle activation. One such trial in Brazil demonstrated that a smart insole with pressure sensors enabled real-time feedback for patients working on post-stroke gait recovery, leading to a 30% increase in walking stability over 6 weeks [27]. By monitoring patient performance outside clinical settings, wearable technology extends care beyond the clinic, which is especially beneficial in LMICs where follow-up opportunities may be rare.
- Robotic and exoskeleton-assisted rehabilitation: Studies on robotic-assisted therapy indicate that robotic exoskeletons can reduce the burden on therapists and enable intensive, repetitive training. A study in China examining the use of a simplified lower-limb exoskeleton found that it significantly improved ambulation in rural patients, allowing them to practice intensive walking therapy without the need for constant supervision [28]. For LMICs, the development of affordable robotic devices with simplified interfaces is critical for wider implementation [29].
- VR and AR for functional recovery: Systematic reviews of VR-based rehabilitation in stroke care report that it enhances both motor skills and patient motivation. In Malaysia, a VR-based intervention designed to improve upper-limb function in stroke survivors showed that patients were more engaged and reported higher satisfaction compared to conventional therapy alone [30]. This promising data suggests that VR, particularly when using portable headsets and gamified exercises, could play an essential role in making rehabilitation more accessible and appealing in LMICs.

1.4.3 Contextualization of Innovations for Low- and Middle-Income Countries (LMICs)

While the effectiveness of these novel approaches is promising, their adaptation for LMICs requires a strategic, context-sensitive approach. Factors such as cost, infrastructure, cultural compatibility, and local healthcare capacity must be considered for successful integration.

1. Cost and scalability: High costs are a significant barrier to the widespread adoption of advanced technologies in LMICs. Innovations like low-cost VR headsets, open-source tele-rehabilitation platforms, and simplified wearable devices are more feasible options for low-resource settings. Collaboration with local technology providers and NGOs could further reduce costs and promote large-scale implementation. For example, VR technology has been scaled down to cost-effective models that can operate on smartphones, making it accessible in settings with minimal infrastructure [22].
2. Training and local capacity building: The integration of these technologies into LMICs requires substantial investment in training for both health professionals and community members. Training rehabilitation staff to manage, maintain, and operate devices such as wearables and robotic aids is essential for sustainable implementation. Additionally, engaging community health workers can facilitate basic rehabilitation in areas lacking specialized personnel, especially for tele-rehabilitation and mobile health services [31].
3. Infrastructure considerations: The digital divide in LMICs poses challenges for tele-rehabilitation and wearable technology, which often rely on stable internet or cellular networks. Innovations such as offline-capable mHealth applications and solar-powered devices have shown promise in mitigating these challenges. In settings with limited internet, mHealth platforms that use SMS-based protocols or offline data collection offer viable alternatives, bridging the technology gap while still enabling effective remote rehabilitation [32].
4. Cultural adaptation and patient engagement: The success of any rehabilitative innovation hinges on patient engagement, particularly when novel technologies are introduced. Adapting technologies to align with local cultural norms and preferences increases patient adherence and long-term success. In Nigeria, for instance, culturally adapted tele-rehabilitation programs that incorporate local language support and consider family involvement have shown high patient engagement and satisfaction [33].

In summary, recent innovations in stroke rehabilitation, encompassing tele-rehabilitation, wearable technology, robotics, and VR, have redefined the possibilities for post-stroke recovery, especially within low-resource settings. Research underscores their efficacy in improving functional outcomes and quality of life, while adapted models provide promising frameworks for LMICs. With continued focus on scalable, affordable adaptations, these technologies hold the potential to bridge the rehabilitation gap, offering stroke survivors in LMICs improved access to recovery pathways and better long-term outcomes.

1.5 Actionable Frameworks for Implementation

The escalating pace of healthcare innovation and the ethical imperative for professionals to deliver effective care underscore the critical role of research evidence in guiding clinical decision-making [34]. Implementing evidence-based practice (EBP) offers substantial benefits for clinicians, patients, and healthcare professionals. Scholars in implementation science frequently emphasize the significant delay between the development of new practices and their adoption, with only 14% of these discoveries ultimately integrated into routine clinical care [35, 36]. This gap highlights the inefficiency of translating scientific discoveries into practical healthcare applications, pointing to a need for improved methods and strategies to accelerate the adoption of evidence-based interventions in clinical settings. Nevertheless, integrating EBP into clinical settings is a multifaceted and ongoing process that necessitates a comprehensive approach encompassing individual, organizational, and patient-centered considerations.

1.5.1 Strategies for Integrating Evidence-Based Practices

Broadly speaking, implementation strategies can be categorized into three approaches:

(a) Passive strategies involve the distribution of educational resources such as posters, tool kits, and visual aids, requiring minimal interaction from recipients.
(b) Active strategies are more engaging and interactive, utilizing workshops, academic detailing, audits, feedback, and reminders to encourage participation and the application of knowledge.
(c) Multimodal translational strategies focus on implementing evidence-based practices by considering the type of knowledge, the context of implementation, and the processes and individuals involved to ensure tailored and effective adoption.

Multimodal strategies did not consistently lead to the actual implementation of EBPs in clinical settings. Conversely, audit and feedback strategies and organizational initiatives designed to increase practice guideline adoption were successful techniques that had a positive impact on the integration of EBPs.

The Expert Recommendations for Implementing Change (ERIC) study aimed to refine an existing collection of implementation strategy terms and definitions by systematically gathering input from a diverse group of stakeholders with expertise in implementation science and clinical practice. The expert panel reached a consensus on a final compilation of 73 implementation strategies. (A refined compilation of implementation strategies: results from the Expert Recommendations for Implementing Change (ERIC) project | Implementation Science (https://link.springer.com).)

1.5.2 Practical Frameworks Tailored for Resource Constraints

Numerous interventions have been proposed to enhance functional, activity, and participation recovery after stroke. However, in many low- and middle-income countries (LMICs), the implementation of these interventions is often hindered by a shortage of qualified rehabilitation specialists and centeres [37, 38]. This scarcity makes the use of resource-intensive rehabilitation strategies unsustainable and impractical.

To address the challenges low-cost robotic therapies in stroke rehabilitation and its potential to alleviate the labor-intensive demands of conventional therapies in low-resource settings are being developed and evaluated for their efficacy. Many of these devices feature one or two degrees of freedom and utilize tendon-driven mechanisms, soft gloves, and off-the-shelf components to replicate hand movements. To minimize costs and enhance portability, traditional materials like steel and metal alloys are avoided. Instead, locally sourced components such as joysticks, splints, hobby servomotors, wood, aluminum alloys, PVC, and plastic are employed [29]. The rise of affordable 3D printing technology has further enabled the use of materials like acrylonitrile butadiene styrene (ABS) and polylactic acid (PLA) in the construction of these devices.

Furthermore, in resource-limited settings, providing stroke rehabilitation necessitates a concerted effort from all stakeholders, including governments, policymakers, rehabilitation professionals, nongovernmental organizations, and foreign collaborators, to train and employ sufficient rehabilitation personnel. To ensure the effectiveness of conventional stroke therapies in LMICs, standardized rehabilitation curricula and clinical practice guidelines are crucial. Additionally, task shifting, the delegation of rehabilitation tasks to less costly non-health workers, merits exploration as a sustainable strategy. However, trials such as the FAmily-Led RehabiliTaTion aftEr Stroke in INDia (ATTEND) and Caregiver-Delivered Stroke Rehabilitation in Rural China (RECOVER) studies reveal insufficient evidence to support its widespread implementation in stroke care, suggesting the need for further research to identify effective models, including potential training for healthcare assistants. Yet, a study in Malawi, where trained community health workers (CHWs) were used with technology assistance for stroke rehabilitation, facilitated patients' mobility and functional capacity in a rural setting.

Tailoring standardized protocols facilitates greater engagement among stroke survivors and promotes a more effective and sustainable rehabilitation process, such as Modified Constraint-Induced Movement Therapy (mCIMT) for stroke patients, which shortens both the intensive training session of the paretic upper extremity to a duration of 30 min to 2 h/day and the restraint time of the nonparetic upper extremity (≤6 h/day) [39]. This intervention can be conducted in a home setting, with active involvement from caregivers or family members. Furthermore, integrating daily activities into the practice regimen can enhance adherence and optimize therapeutic outcomes.

In addition to these strategies, a greater focus should be placed on structured rehabilitation programs that are simple, scalable, and adaptable to low-resource contexts. For instance, the Graded Repetitive Arm Supplementary Program (GRASP https://neurorehab.med.ubc.ca/grasp/) is a self-directed, upper-limb exercise program that relies on basic, inexpensive materials and can be integrated into both inpatient and community settings with minimal supervision. Similarly, the Patient-Led Upper-Limb Strengthening at Home (PUSH https://strokeed.com/wp-content/uploads/2020/01/PUSH-UL-program-Aug-2011.pdf) program enables stroke survivors to continue progressive upper-limb strengthening independently, supporting continuity of care beyond the clinic. Meanwhile, the Fitness and Mobility Exercise (FAME https://fameexercise.com/) program offers a low-cost, group-based intervention targeting strength, balance, mobility, and cardiovascular fitness, making it well suited for community or outpatient delivery. Lookout for the work on the roundtable on High dose high intensity training at home by the International Stroke Recovery and Rehabilitation Alliance (ISRRA) in 2026.

Lastly, tele-rehabilitation addresses access barriers by offering remote services, with emerging evidence supporting its effectiveness in improving quality of life and reducing caregiver stress [25]. Smartphone-based interventions offer a promising avenue for education and support, with rising ownership rates (around 45% in emerging economies) enabling initiatives like the "Care for Stroke" app in India, which provides instructional resources for caregivers and stroke survivors post-discharge. The F@ce program in Uganda exemplifies the effective use of SMS to enhance participation and goal-setting among patients and their families. For areas lacking smartphone access, video-based instruction provides an alternative method of delivering rehabilitation. Trials in Thailand and Malaysia have demonstrated that audiovisual materials, paired with home visits from physical therapists, can yield positive outcomes in activities of daily living and mental health for stroke survivors.

1.5.3 Recommendations for Policymakers and Stakeholders

1.5.3.1 Educational Policy

The absence of a centralized regulatory body for rehabilitation in many LMICs often results in disparate practices across different regions within the same nation. Consequently, the establishment of a national rehabilitation council, equipped with a standardized curriculum, is recommended. In the interim, where such a council is not yet established, the Ministry of Health (MOH) could proactively form a task force to develop a comprehensive and scalable stroke rehabilitation education program. This program should encompass basic to advanced levels, emphasizing practical implementation strategies and tools. By providing both students and practicing therapists with evidence-based training, this initiative aims to enhance patient outcomes and ensure consistent, high-quality rehabilitation services across the country.

1.5.3.2 Development and Implementation of Guidelines

The reliance on poorly established practices is more pronounced in regions where there is a lack of appropriate guidelines or where existing guidelines are not suitable for the local context [40, 41]. Therefore, developing national guidelines that are evidence-based and tailored to the specific needs of the country could significantly enhance rehabilitation services [26].

1.5.3.3 Access to Licenced Rehabilitation Services

To ensure the quality and accountability of rehabilitation services, policies should mandate that rehabilitation providers and centers are licensed by the country's health ministry. This licensing process should include rigorous evaluation of qualifications, experience, and adherence to established standards of practice. Additionally, the government should conduct regular audits of rehabilitation centers to ensure quality and compliance with the key quality indicators.

1.5.3.4 Insurance for Rehab

Low- and middle-income countries (LMIC) rely significantly on private health expenditure in the form of out-of-pocket payments (OOP) and voluntary health insurance (VHI). Most VHIs do not provide coverage for rehabilitation services and hence rely on substandard services . To ensure equitable access to high-quality rehabilitation care, it is essential to implement policies that mandate the inclusion of rehabilitation services in national health insurance plans or other publicly funded healthcare programs [18, 42].

1.5.3.5 Community-Based Rehab Programs

The seamless integration of rehabilitation services across all levels of care, from primary to quaternary, is essential. Currently, there is a bias toward higher levels of care, but well-resourced community rehabilitation programs can offer cost-effective, patient-centered solutions. Local stroke champions and rehabilitation teams should lead these programs, which can be implemented in community halls or similar venues to reduce infrastructure costs [43].

Additionally, public–private partnerships should be encouraged to expand community-based rehabilitation services and raise awareness through support groups, ensuring more widespread access and education on the benefits of rehabilitation.

What Does Evidence Say

- Access to an interprofessional team that can provide recommended amount of rehabilitation is a must for every stroke survivor.
- An initial assessment (within 48 h of stroke onset) by a multidisciplinary team is recommended for those admitted with acute stroke.
- Consider the use of telehealth for those who have demonstrated benefit either as an alternative or adjunct to in-person rehabilitation (where regular in-person sessions may be difficult).
- Rehabilitation goals should be identified collaboratively with patients and caregivers, with clear documentation and provisions for regular reviews throughout the care continuum.
- Caregivers of stroke survivors should be empowered with tailored information and support during the recovery process. Such support could include information, referrals to relevant health professionals, community services in a mode that suits each person the best.
- Within the first 4 months after discharge, every stroke survivor should be prescribed self-directed management interventions in community living.

References

1. Langhorne P, et al. Stroke rehabilitation: international best practices. Lancet Neurol. 2020;19(7):641–53.
2. Winstein CJ, et al. Guidelines for adult stroke rehabilitation and recovery. Stroke. 2016;47(6):e98–e169.
3. Daly JJ, Ruff RL. Evidence-based physical rehabilitation approaches for improving stroke recovery. J Stroke Res. 2022;17(3):190–205.
4. Nair T, et al. Implementing tele-rehabilitation for stroke in rural India. Indian J Stroke Rehabil. 2022;12(3):105–12.
5. Moyo M, et al. Task-shifting in stroke rehabilitation: a low-resource case in Kenya. J Commun Health Rehabil. 2020;5(2):101–11.
6. Grove R, et al. Feasibility of CIMT using local materials in Mozambique. Global Health Rehabil. 2021;9(1):52–61.
7. Kabogo R, et al. Community health worker-led rehabilitation for post-stroke recovery in rural Malawi. Afr J Neurol. 2023;8(4):210–20.
8. Cabral N, Muller M, Franco S, Longo A, Moro C, Nagel V, Liberato R, Garcia A, Venancio V, Gonçalves A. Three-year survival and recurrence after first-ever stroke: the Joinville stroke registry. BMC Neurol. 2015;15:70.
9. Taberna M, Gil Moncayo F, Jané-Salas E, Antonio M, Arribas L, Vilajosana E, Peralvez Torres E, Mesía R. The multidisciplinary team (MDT) approach and quality of care. Front Oncol. 2020;10:85.
10. Jessup RL. Interdisciplinary versus multidisciplinary care teams: do we understand the difference? Aust Health Rev. 2007;31(3):330–1.

11. Lindsay P, Furie KL, Davis SM, Donnan GA, Norrving B. World stroke organization global stroke services guidelines and action plan. Int J Stroke. 2014;9:4–13.
12. Kanwal S, Hashmi R, Perveen W, Ali MA, Akhtar M, Munawar A. Outcomes of multidisciplinary team approach in rehabilitation of patients with stroke. Pak J Med Health Sci. 2022;16(3):78. https://doi.org/10.53350/pjmhs2216378.
13. Deng Y, Sang Y, Shang Y, Wu C, Xu X. To explore the application value of nursing staff involved multidisciplinary continuous nursing in stroke patients with limb dysfunction. BMC Health Serv Res. 2024;24(1):1051.
14. Rathore FA, Anwar F, Younas U. Multidisciplinary team working in rehabilitation medicine: advantages and challenges. J Pak Med Assoc. 2024;74(2):409–12.
15. Adeniji O, Adeleye O, Akinyemi J, Otubogun F, Ogunde G, Ogunrombi M, Adesina D, Wahab A, Ogunlana M, Alimi T, Akinyemi R. Organized multi-disciplinary stroke team care improves acute stroke outcomes in resource limited settings; results of a retrospective study from a Nigerian tertiary hospital. J Stroke Cerebrovasc Dis. 2023;32(10):107307. https://doi.org/10.1016/j.jstrokecerebrovasdis.2023.107307.
16. Gopaul U, et al. Arch Phys Med Rehabil. 2023;104(10):1745–50.
17. An Z, Li K, Yang X, et al. Community-based rehabilitation services implemented by multidisciplinary teams among adults with stroke: a scoping review with a focus on Chinese experience. BMC Public Health. 2024;24:740. https://doi.org/10.1186/s12889-024-18218-1.
18. Bernhardt J, Urimubenshi G, Gandhi DBC, Eng JJ. Stroke rehabilitation in low-income and middle-income countries: a call to action. Lancet. 2020;396(10260):1452–62. https://doi.org/10.1016/S0140-6736(20)31313-1.
19. Cramer SC, et al. Digital therapeutics in stroke recovery. Stroke. 2019;50(12):3335–43.
20. Liao W, et al. Wearable technology for post-stroke gait analysis and intervention. Gait Posture. 2021;85:135–42.
21. Díaz R, et al. Robotics in low-resource stroke rehabilitation: feasibility and cost-effectiveness. Int J Neurorehabil. 2022;29(3):200–9.
22. Fung V, et al. Virtual reality in stroke rehabilitation: systematic review and meta-analysis. Stroke Rehabil. 2021;38(4):229–36.
23. Bello ML, et al. Efficacy of tele-rehabilitation in LMICs: a meta-analysis. J Glob Health. 2023;13(2):151–64.
24. Nedjat S, et al. Comparative study on tele-rehabilitation efficacy in South Africa. Afr Neurol. 2022;15(3):102–11.
25. Gandhi DBC, Kamalakannan S, Dsouza JV, Montanaro V, Chawla NS, Mahmood A, Ngeh E, Zarreen S, Vijayanand PJ, Solomon JM. Research, education and practice of tele-neurorehabilitation in low and middle-income countries: a scoping review. NeuroRehabilitation. 2024; https://doi.org/10.3233/NRE-240053.
26. Gandhi DBC, Kamalakannan S, Urimubenshi G, Sebastian IA, Montanaro VVA, Chawla NS, D'Souza JV, Ngeh E, Mahmood A, Demers M, Hombali A, Solomon JM, Collaborators GCSR. Stroke rehabilitation clinical practice guidelines in low- and middle-income countries: a systematic review of quality and unique features. Cerebrovasc Dis. 2024; https://doi.org/10.1159/000539999.
27. Reis TL, et al. Evaluation of smart insole efficacy in post-stroke gait rehabilitation. Latin Am J Biomech. 2021;23(2):145–53.
28. Zhang Y, et al. Simplified exoskeletons for rural stroke recovery: a study in China. Chin J Phys Med Rehabil. 2022;35(2):67–75.
29. Demofonti A, Carpino G, Zollo L, Johnson MJ. Affordable robotics for upper limb stroke rehabilitation in developing countries: a systematic review. IEEE Trans Med Robot Bionics. 2021;3(1):11–20.
30. Nguyen H, et al. Cost-effective VR for post-stroke rehabilitation in low-resource settings. Asia Pacific J Rehabil. 2021;14(3):84–91.
31. Louw Q, Grimmer K, Dizon JM, Machingaidze S, Parker H, Ernstzen D. Building capacity in primary care rehabilitation clinical practice guidelines: a south African initiative. Health Res Policy Syst. 2018;16:1.

32. Gueye A, et al. mHealth interventions in LMICs: overcoming connectivity challenges. Digital Health. 2020;6:34–45.
33. Adeyemi T, et al. Cultural adaptations in tele-rehabilitation for stroke: a Nigerian case study. Afr J Telemed. 2023;8(1):43–51.
34. Balas EA, Boren SA. Managing clinical knowledge for health care improvement. Yearb Med Inform. 2000;9(01):65–70.
35. Morris ZS, Wooding S, Grant J. The answer is 17 years, what is the question: understanding time lags in translational research. J R Soc Med. 2011;104(12):510–20.
36. Goorts K, Dizon J, Milanese S. The effectiveness of implementation strategies for promoting evidence informed interventions in allied healthcare: a systematic review. BMC Health Serv Res. 2021;21(1):241. https://doi.org/10.1186/s12913-021-06190-0.
37. Ekechukwu END, Olowoyo P, Nwankwo KO, et al. Pragmatic solutions for stroke recovery and improved quality of life in low- and middle-income countries-a systematic review. Front Neurol. 2020;11:337. https://doi.org/10.3389/fneur.2020.00337.
38. Juckett LA, Wengerd LR, Faieta J, Griffin CE. Evidence-based practice implementation in stroke rehabilitation: a scoping review of barriers and facilitators. Am J Occup Ther. 2020;74(1):7401205050.
39. Baldwin CR, Harry AJ, Power LJ, Pope KL, Harding KE. Modified constraint-induced movement therapy is a feasible and potentially useful addition to the community rehabilitation tool kit after stroke: a pilot randomised control trial. Aust Occup Ther J. 2018;65(6):503–11.
40. Prvu Bettger J, Liu C, Gandhi DB, Sylaja PN, Jayaram N, Pandian JD. Emerging areas of stroke rehabilitation research in low-and middle-income countries: a scoping review. Stroke. 2019;50(11):3307–13.
41. Lynch EA, Connell LA, Carvalho LB, Bird ML. Do clinical guidelines guide clinical practice in stroke rehabilitation? An international survey of health professionals. Disabil Rehabil. 2022;44(15):4118–25. https://doi.org/10.1080/09638288.2021.1891304.
42. Dee M, Lennon O, O'Sullivan C. A systematic review of physical rehabilitation interventions for stroke in low and lower-middle income countries. Disabil Rehabil. 2020;42(4):473–501. https://doi.org/10.1080/09638288.2018.1501617.
43. Nowrin I, Mehareen J, Bhattacharyya DS, Saif-Ur-Rahman KM. Community-based interventions to prevent stroke in low and middle-income countries: a systematic review. Health Sci Rev. 2023;9:100123.

Post-Stroke Rehabilitation in the Acute Phase

2

Dorcas B. C. Gandhi, Vinicius A. Montanaro, Rinita Mascarenhas, and Abhishek Srivastava

Abstract

Post-stroke recovery involves two distinct yet interconnected aspects: rehabilitation, a structured intervention process led by healthcare professionals, and recovery, driven by neuroplasticity, the brain's ability to reorganize and restore function. The recovery timeline is divided into hyperacute, acute, subacute, and chronic phases, with the early stages (acute and subacute) offering a critical window of heightened neuroplasticity for targeted interventions that maximize functional recovery. Prognostic tools like the PREP-2 and TWIST models, alongside baseline functional assessments, enable clinicians to predict recovery outcomes

D. B. C. Gandhi (✉)
Department of Neurology, Christian Medical College & Hospital Ludhiana, Ludhiana, Punjab, India

Department of Physiotherapy, Manipal Academy of Higher Education, Manipal, Karnataka, India
e-mail: dorcas.gandhi@cmcludhiana.in

V. A. Montanaro
Neurological Rehabilitation SARAH Network of Rehabilitation Hospitals, Brasília, Brazil

Radbou University Medical Center, Nijmegen, Netherlands

Programa de Pós Grduação em Neurologia da Universidade Federal Fluminense, Niterói, Brazil

SARAH Network of Rehabilitation Hospitals, Brasília, Brazil

R. Mascarenhas
Department of Neurology, Christian Medical College and Hospital, Ludhiana, Punjab, India

A. Srivastava
Center for Physical Medicine and Rehabilitation, Kokilaben Dhirubhai Ambani Hospital and Medical Research Institute, Mumbai, India

D. B. C. Gandhi, J. D Pandian (eds.), *Stroke Rehabilitation in Low-Resource Settings*, https://doi.org/10.1007/978-981-92-0403-8_2

and tailor rehabilitation plans, particularly in resource-limited settings. Acute-phase rehabilitation focuses on early mobilization, task-specific training, and the prevention of complications, with neurologists stabilizing patients and physiatrists initiating functional recovery efforts through individualized therapy regimens. In low-resource settings, innovative strategies like task-shifting, telemedicine, and standardized care pathways bridge systemic gaps, enabling effective and accessible rehabilitation for stroke survivors.

Keywords

Acute stroke rehabilitation • Early mobilization • Screening and assessment • Complication prevention • Risk management

2.1 Introduction

2.1.1 Understanding Post-Stroke Recovery

The World Health Organization (WHO) defines stroke rehabilitation as the coordinated efforts of two or more healthcare disciplines working alongside medical professionals to deliver targeted interventions. Rehabilitation and recovery are often used interchangeably. However, rehabilitation refers to the structured process of intervention, while recovery describes the mechanisms by which improvement occurs, whether through the restoration of lost functions or the adoption of compensatory strategies.

Recovery after a stroke is primarily driven by neuroplasticity, the brain's ability to adapt and reorganize by inducing functional changes in both the affected (ipsilesional) and unaffected (contralesional) regions. It refers to the nervous system's ability to modify its structure and function in response to internal and external stimuli. These changes facilitate the restoration of motor and other system functions.

A significant portion of early improvement occurs through spontaneous recovery (SR). SR refers to natural, therapy-independent improvements in behavior or function, such as motor skills. SR typically begins shortly after a stroke, peaks around 4 weeks, and tapers off over 6 months, with variations depending on the system affected. Early recovery (within the first 3–6 months) often results from spontaneous mechanisms, leading to improvements in motor control, language, and other primary neurological functions.

In contrast, true recovery (TR), also known as behavioral restitution, involves a return to pre-stroke patterns of motor control, which requires neural repair and the restoration of original function. Recovery may also occur through compensation, where patients achieve functional goals by adopting new strategies and relying on intact muscles and joints. This process focuses on skill relearning rather than neural repair and often underpins rehabilitation strategies, enabling individuals to achieve functional independence despite physical limitations. Later recovery, marked by an improved ability to perform daily activities despite physical impairments, also stems from neuroplasticity. This adaptability can be enhanced through interventions like pharmacological agents, electrical stimulation, task-specific training, and environmental modifications, fostering further functional gains beyond the initial recovery phase.

2.1.2 Timeline of Recovery and the Importance of Initiating Rehabilitation in the Acute Phase

Building on prior research by Dobkin and Carmichael, a conceptual framework drawn from preclinical studies using animal stroke models, as well as clinical research involving stroke survivors, particularly in the domain of motor system recovery, was created by the International Stroke Recovery and Rehabilitation Alliance [1]. It remains a dynamic model, subject to updates as discoveries emerge. This framework highlights the timeline (spanning hours to months) of key biological processes following ischemic and hemorrhagic stroke. It also maps the commonly used temporal phases of recovery: hyperacute, acute, early, and late subacute, and chronic across the first 6 months and beyond. While behavioral improvements can occur years after a stroke, current evidence suggests that the most significant and rapid recovery tends to occur within the initial weeks to months post-stroke for most individuals. This period, marked by heightened neuroplasticity and brain repair processes, represents a critical window for therapeutic interventions aimed at maximizing functional recovery [1, 2]. Early and intensive inpatient rehabilitation during the acute and subacute phases is strongly recommended, as evidence shows that it improves functional outcomes. Systematic reviews and meta-analyses highlight its role in reducing mortality and promoting independence in stroke units.

2.1.3 Rehabilitation-Related Prognostic Factors in the Acute Phase Post-Stroke

A well-known fact is that baseline ADL function is a key predictor of post-stroke independence, with limitations in sitting and hemiplegic lower limb motor impairment serving as significant prognostic factors [3]. Such measures can easily be performed even in low-resource settings to develop customized therapy plans for patients. Emerging predictive models offer clinicians valuable tools to estimate post-stroke recovery outcomes more efficiently and accurately. One notable example is the Predict Recovery Potential (PREP-2) algorithm, an improved version of the original PREP-1. PREP-2 utilizes the SAFE score (Shoulder Abduction and Finger Extension muscle strength grades) within the first 3 days post-stroke, along with patient age, NIHSS score, and the presence of Motor Evoked Potentials (MEPs), to predict four potential upper limb outcomes at 3 months. CM Stinear et al. also outlined tailored rehabilitation strategies for each predicted outcome. PREP-2 achieves up to 75% accuracy in its predictions. In low-resource settings where MEP cannot be used as standard measures, SAFE score values can be beneficial for planning rehabilitation regimens for the UL in the acute and early subacute phases [4].

For walking ability, the "Time to Walk Independently after Stroke" (TWIST) model offers reliable predictions using the Trunk Control Test to classify walking potential based on the Functional Ambulatory Category (FAC), with an accuracy of 91–100%. This prognostic tool also needs limited or no resources other than a trained workforce [5]. However, it is essential to emphasize that these models provide probabilistic outcomes and rely heavily on factors such as exercise intensity, patient motivation, and adherence to therapy for optimal results.

2.2 Medical Aspects of Acute Post-Stroke Rehabilitation

Early intervention in the acute phase, typically within the first 2–14 days post-stroke, is critical for optimizing functional outcomes and preventing complications. Neurologists and physiatrists are at the forefront of this effort, working to stabilize patients, manage neurological deficits, and initiate early rehabilitation [6, 7]. Addressing the unique challenges of stroke rehabilitation in LMICs requires evidence-based approaches tailored to resource-limited settings [8].

2.2.1 Medical Priorities in Acute Phase

Neurological Stabilization and Acute Management: Neurologists play a pivotal role in acute-phase stroke care, focusing on rapid diagnosis, medical stabilization, and prevention of secondary injury.

Rapid Diagnosis and Early Intervention: Accurate differentiation between ischemic and hemorrhagic strokes is critical, as treatment strategies differ significantly. In LMICs, limited access to advanced imaging (e.g., CT or MRI) necessitates reliance on clinical tools such as the National Institutes of Health Stroke Scale (NIHSS) and the Glasgow Coma Scale (GCS) for initial assessment [6, 8]. For ischemic strokes, early administration of intravenous tissue plasminogen activator (tPA) within 4.5 h can significantly reduce disability. However, logistical barriers often delay this treatment in resource-limited settings [9]. Efforts to simplify protocols and improve prehospital care pathways are essential for broader tPA utilization.

Managing Blood Pressure and Preventing Complications: Rapid and frequent blood pressure (BP) monitoring is a core component of stroke care. In hemorrhagic strokes, rapid blood pressure control is critical to prevent hematoma expansion. The Intensive Blood Pressure Reduction in Acute Cerebral Hemorrhage Trial demonstrated that maintaining systolic blood pressure below 140 mmHg significantly reduces mortality and disability. In ischemic stroke, BP has to be monitored pre- and post-thrombolysis to minimize the risk of intracerebral hemorrhage. If the patient is off the window period for thrombolysis, BP monitoring is essential. However, achieving this target in LMICs requires affordable antihypertensive medications and consistent monitoring, which is often challenging in under-resourced settings.

Neurologists also focus on preventing complications such as aspiration pneumonia, deep vein thrombosis (DVT), and pressure ulcers. Implementing protocols for early mobilization, dysphagia screening, and DVT prophylaxis is crucial. In LMICs, where resources are limited, simple interventions, such as proper patient positioning and early initiation of physical therapy, can significantly reduce these risks [6].

2.2.2 Physiatrist's Role: Early Rehabilitation and Functional Restoration

Physiatrists complement neurologists by focusing on functional recovery and complication prevention through early rehabilitation. Early initiation of rehabilitation, even within the first 24–48 h, can significantly improve outcomes by promoting neuroplasticity and preventing secondary complications [6]. The rehabilitation program can be divided into therapeutic interventions for five major issues: continuation of medical–physiological function, prevention of secondary complications affecting rehabilitation, initiation of training for sensorimotor and functional deficits, training for cognitive–communication–oromotor deficits, and managing psychological and social issues [9].

2.2.3 Comprehensive Neurological Assessment

A thorough assessment of motor, sensory, cognitive, and swallowing functions is the foundation of individualized rehabilitation plans. Tools such as the Modified Rankin Scale (mRS) and the Barthel Index provide standardized measures of functional impairment and recovery progress [6].

2.2.4 Early Mobilization and Task-Specific Training

Evidence from the AVERT trial highlighted that the benefits of early mobilization (first 48 hours) in reducing disability, improving functional outcomes, and reducing complications like pneumonia and DVT, while enhancing motor recovery. On the contrary, very early mobilization, within 24 hours post stroke increases the odds of death and dependency. Readers should also watch for the results paper of the AVERT-DOSE trial, which are intended to clarify the optimal regimen for mobility-focused rehabilitation through out-of-bed mobilisation.

In LMICs, where access to specialized therapists is limited, task-shifting programs that train nurses or community health workers to perform basic rehabilitation interventions can effectively extend services [10]. Task-specific training, which involves practicing activities of daily living (ADLs), has shown promise in improving functional independence.

Spasticity and tone management: Spasticity can develop early in the post-stroke period, especially in patients with severe motor deficits. Effective management often involves pharmacological interventions such as baclofen or botulinum toxin. However, in resource-limited settings, cost-effective alternatives are vital [6].

2.2.5 Addressing LMIC Challenges with Context-Specific Solutions

The acute rehabilitation phase in LMICs faces systemic challenges, including inadequate infrastructure, a shortage of specialized personnel, and financial constraints. Addressing these barriers requires innovative, scalable solutions.

Task-shifting models: Task-shifting involves training nonspecialist healthcare providers, such as nurses or community health workers, to deliver basic rehabilitation services under remote supervision. Programs in countries like Ghana and Uganda have demonstrated that, with proper training, nonspecialists can effectively implement rehabilitation protocols, improving patient outcomes [11].

Telemedicine and tele-rehabilitation: Telemedicine offers a promising solution to the shortage of specialists in LMICs, as trained specialists can provide remote consultations, guiding local providers in managing acute stroke and initiating rehabilitation. Studies have shown that tele-rehabilitation can improve functional outcomes and patient satisfaction, even in low-resource settings [12, 13].

Standardized care pathways: Implementing standardized protocols for acute stroke management ensures consistency and adherence to evidence-based practices. Guidelines adapted to local resources can include simplified assessment tools, early mobilization protocols, and essential secondary prevention strategies [7, 8].

2.3 Rehabilitation in the Acute Phase Post-Stroke

2.3.1 Key Objectives of Acute Phase Rehabilitation

The main objectives of rehabilitation in the acute phase post-stroke are: (1) prevent complications (e.g., contractures, pressure ulcers), (2) facilitate early mobilization, and (3) lay the foundation for functional recovery. A structured assessment and intervention plan is essential in the acute phase within a timeframe deemed effective by clinical practice guidelines [1].

Early assessments and predictions should include assessment and intervention for out-of-bed mobility within the first 48 h of onset; however, too early initiation of such mobility tasks within 24 h has been reported to negatively affect death and dependency on mRS at 3 and 6 months post-stroke.

2.3.2 Key Areas of Assessment

Timely assessments in the acute phase are highly essential to predict recovery and plan suitable and customized therapy regimens. This could include, but is not limited to, recommended neurological assessments with the National Institutes of Health Stroke Scale (NIHSS) and Glasgow Coma Scale (GCS), functional assessments with the modified Rankin Scale (mRS) and modified Barthel Index (mBI), sensorimotor impairment assessment with the Fugl-Meyer Assessment (FMA) within the first 48 h of stroke, cognitive and communication screening with Montreal

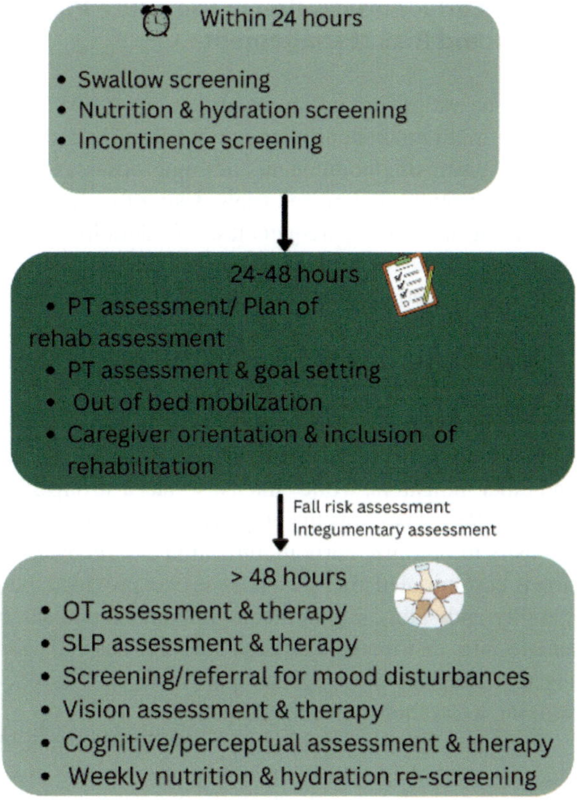

Fig. 2.1 Timeline of essential assessments in post-stroke acute phase

Cognitive Assessment (MoCA) in the first week post-stroke, swallowing and nutritional/hydration status with bedside swallowing assessment, or videofluoroscopy within 24 h of stroke onset (Fig. 2.1).

Screening for dysphagia in stroke patients should be conducted as early as possible, beginning with simple bedside tests like water-swallowing tests or more comprehensive multi-consistency protocols. For some patients, a detailed evaluation using instrumental diagnostics, such as flexible endoscopic swallowing evaluation or videofluoroscopy, is necessary. Emerging diagnostic methods, advancements in assessment technologies, and digital innovations are expected to enhance diagnostic precision in the future [14].

Improvements in dysphagia diagnosis will support individualized management strategies tailored to specific dysfunction patterns and risk factors for complications. Progress in dysphagia rehabilitation is critical to reducing mortality and enhancing the quality of life for stroke survivors. The GUSS (Gugging Swallowing Screen) provides a fast and reliable tool for identifying stroke patients at risk of dysphagia and aspiration. This graded assessment takes a more nuanced approach to evaluating the pathophysiology of voluntary swallowing, minimizing patient discomfort. It allows those who can safely consume semisolid foods to continue their oral feeding routine while avoiding fluids that may pose a higher risk. For further details, refer to Chap. 6.

2.3.3 Early Mobilization in the Acute Phase and Risk Management

Need: The multiphase optimization strategy (MOST) offers a structured approach to refining multicomponent interventions through a three-stage process: screening to identify promising components, refining doses and combinations, and confirming benefits through randomized trials. Using this framework, we explored the effects of varying intensities, frequencies, and durations of early mobilization on stroke recovery. Researchers report that early mobilization should begin within 24–48 h post-stroke, incorporating activities like bed-to-chair transfers, sitting, standing, walking, or climbing stairs, with intensity tailored to the patient's ability while avoiding fatigue. High-frequency mobilization may lead to increased fatigue, negatively impacting self-efficacy and social participation. Researchers also suggest that the "frequency" of mobility sessions is the most significant factor influencing outcomes such as the Barthel Index (BI), Stroke-Specific Quality of Life (SS-QoL), and other indicators, while "intensity" has a stronger effect on the modified Rankin Scale (mRS). These results emphasize the importance of balancing frequency and intensity to optimize early mobilization and recovery. In summary, early mobilization is best initiated after 24 h post-stroke, provided the patient meets hemodynamic stability and safety requirements. Recommended sessions should last between 15 and 45 min, distributed across one to three sessions per day. The primary goal during this phase is to engage the patient in activities such as sitting, standing, and walking to promote recovery [15].

Prerequisites: Early post-stroke mobilization should be planned carefully, considering several factors to ensure safety and effectiveness. Even in low-resource settings with simple monitoring criteria, a safe and feasible plan can be made. One example is the process followed in India, where the timing of mobilization depends on the type and severity of the stroke. Patients with severe hemorrhage involving ventricular breakthrough, significant cerebral edema, or other serious complications should not undergo early mobilization even within 48 h. Favorable clinical parameters include a systolic blood pressure (SBP) range of 120–210 mmHg for ischemic strokes, 120–160 mmHg for hemorrhagic strokes, and 120–185 mmHg for thrombotic strokes. Additionally, heart rates should remain between 50 and 120 beats/min, oxygen saturation should be maintained at 93% or higher, and stroke symptoms should be stable.

The NICE guidelines for early mobility outline specific prerequisites regarding vital signs before initiating a mobility session. They recommend transitioning the patient to a chair only if they can tolerate sitting at the edge of the bed with assistance from 2 to 3 therapists and, if needed, additional head support. Mobility should be avoided in unconscious patients but can be initiated in drowsy patients under careful supervision [16].

Monitoring: Out-of-bed mobility is encouraged for patients who can maintain alignment while sitting with the support of 1–2 therapists, are alert, and respond appropriately to commands. Adequate staffing is essential to ensure patient safety, particularly in the absence of a tilt table. Monitoring during mobilization is essential to avoid adverse events. Blood pressure should be checked before initiating the session, after 3–5 min of sitting at the edge of the bed, and within 20 min of starting the mobility exercises. In cases where BP monitoring is not feasible, signs such as fatigue, restlessness, facial flushing, excessive perspiration, breathlessness, dizziness, drowsiness, nausea, or vomiting should be closely observed. Mobilization should be stopped immediately if SBP drops by more than 20 mmHg or if any negative clinical signs are observed. If two or more such issues arise, mobilization should be paused for at least 3 h before attempting the next session, with careful monitoring during subsequent sessions. Monitoring can be discontinued after two successful sessions without any adverse events. This structured approach ensures that mobilization is safe, effective, and tailored to the patient's clinical condition, thereby optimizing recovery outcomes.

Content of the mobility sessions: The content of mobilization sessions should be tailored to the patient's condition and stroke severity. For severe strokes, activities may focus on bed mobility, including trunk and pelvic girdle movements, turning in bed, and closely monitored bed-to-chair transfers. In contrast, patients with stable strokes and higher sensorimotor function, as indicated by the Fugl-Meyer Assessment (FMA), may progress to advanced tasks such as stair climbing, balance training, and dynamic posture exercises. Intermediate-level activities can include repetitive upper limb and trunk tasks while seated at the edge of the bed, supervised walking within the ICU or wards, spot marching, and stepping movements (forward, backward, or sideways) with or without support. Standing upper limb exercises may also be incorporated. Contextualized exercises across FMA severity categories, along with details of delivery personnel and training considerations in LMIC settings, are summarized in (Table 2.1).

To enhance engagement and performance, auditory or visual feedback can be used, and maintaining live conversation and constant monitoring are crucial. For patients with mild to moderate strokes, as assessed by the NIH Stroke Scale (NIHSS), sessions can be longer and more frequent. However, for those with severe strokes, particularly hemorrhagic types, sessions should be shorter and less frequent to avoid fatigue and ensure safety. This individualized approach maximizes recovery while minimizing risks.

2.3.4 Introduction of Occupational Therapy Therapies in the Acute Phase

The focus of occupational therapy in the acute phase includes enhancing functional independence in activities of daily living (ADLs) and preventing learned nonuse of the affected side. Evidence-based interventions emphasize the early initiation of ADLs, such as grooming, dressing, and feeding, to promote independence and recovery. Task-specific training is a critical component for improving upper-limb function, while cognitive stimulation exercises are essential for patients with cognitive impairments to enhance mental engagement and overall recovery. In low-resource settings, contextualized strategies can ensure the continuity and

Table 2.1 Contextualized exercises for the four categories of FMA, along with a column on who/how can deliver it (with training) in LMICs

FMA categories	Domain	Exercise examples (not exhaustive)	Contextualization and tips for implementation	Who/how can deliver in LMICs (with training)
0–35 (very severe)	Bed mobility	• Log and segmental rolling • Head-up positioning with bilateral arm activities • Supported sitting with cushions/pillows • Mirror therapy/action observation therapy	• Train caregivers or on-call nurses • Combine with nursing sessions (hygiene, bedsheet change, feeding) • Use low-cost aids (pillows, towels, sandbags)	• Include caregivers in the session • Nursing staff during daily care
36–55 (severe)	Basic functional tasks	• Sit-to-stand with support • Tabletop activities (supported reaching for household objects—Cups, bottles, towels) • Weight shifting in sitting and supported standing • Assisted stepping • Early out-of-bed mobilization	• Use household objects (cups, bottles, towels) for reaching tasks • Family/caregivers can supervise short bouts • Focus on repetition in safe, supervised environments	• Family/caregivers for supervised practice
56–79 (moderate)	General ADLs and trunk engagement	• ADL training: self-feeding, grooming, dressing practice • Reaching across midline (picking from floor or opposite side) • sTep-ups, mini-squats, sit-to-stand without support • Endurance: cycle ergometer, walking short distances	• Use locally available adaptive aids (thick-handled spoons, elastic bands) • Combine with family/caregiver training for multiple short sessions • Integrate into daily household chores (table clearing, folding clothes)	• Physiotherapists/OTs • Nursing staff • Family members for daily supervised practice
>79 (mild)	Advanced ADLs and community reintegration	• Functional tasks: chopping with butter/wooden knife, stirring, carrying pots, full laundry cycle (washing, wringing, hanging, folding) • Circuit training: balance and endurance • Dual-tasking: conversation while walking, counting backwards during step-ups	• Integrate therapy into real-life home/community tasks • Create a list of activities based on the prioritization of patients' needs	• Patient self-practice (with structured guidance)

effectiveness of rehabilitation. Available items can be adapted to create task-specific exercises, enabling patients to practice functional tasks using readily available resources. Additionally, educating family members empowers them to support therapy at home, bridging the gap between professional care and daily practice. These approaches ensure that rehabilitation remains accessible and effective, even in resource-limited environments.

2.3.5 Speech-Language Pathology (SLP) in the Acute Phase

For further details please refer to Chap. 6

2.3.6 Contextualized Regimens for Low-Resource Settings

Acute phase rehabilitation in low-resource settings faces challenges like limited access to skilled therapists, equipment, and facilities. To address this, family caregivers can be trained as therapy aides, and locally available materials can be used to create affordable alternatives like splints and exercise aids. Similar to task shifting in community-based programs, acute phase rehabilitation post-stroke can be improved by leveraging local materials and manpower resources. Success stories demonstrate how adapted protocols and innovative strategies can overcome these limitations and improve patient outcomes effectively. An example is the family-led rehabilitation after stroke trial, which found no significant impact on the modified Rankin Scale (mRS) at 3 months. However, patients and their caregivers reported feeling empowered and safe following regular training sessions conducted both in the hospital and at home. This trial highlights the importance of supervised task-shifting strategies, particularly in low-resource settings, to support effective rehabilitation.

2.4 Immediate Post-Stroke Complications that Need Rehabilitation Prevention and Management

In-hospital complications significantly increase mortality rates among acute stroke patients, both during hospitalization and in the subsequent months. Patients presenting with multiple complications upon admission are at particularly high risk of adverse outcomes. Figure 2.2 illustrates some of the common post-stroke complications that effective rehabilitation can help prevent or mitigate [17].

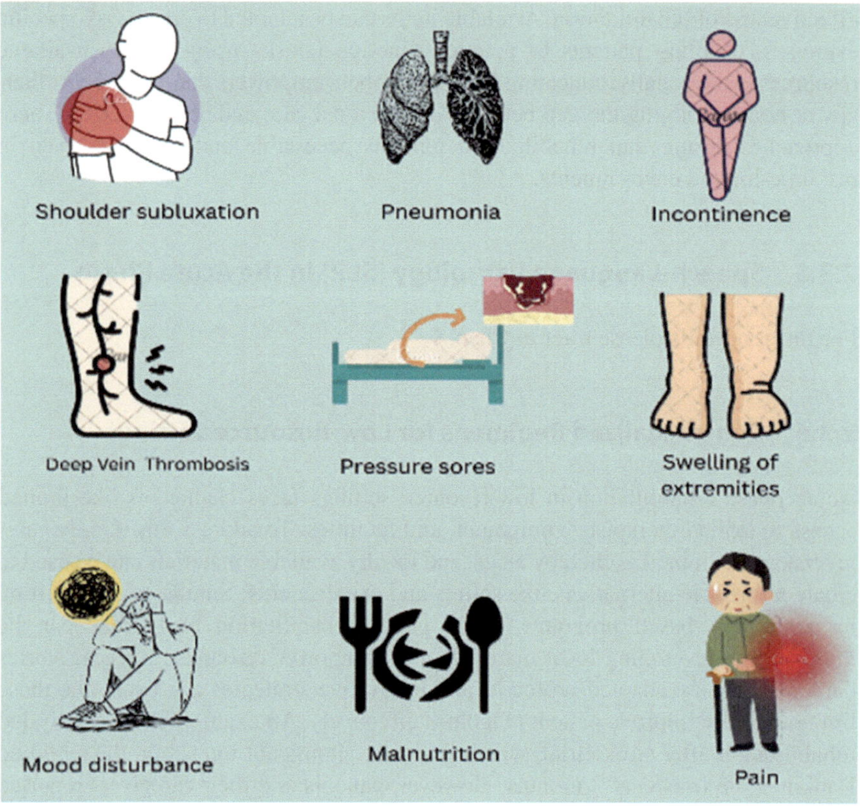

Fig. 2.2 Post stroke complications

2.4.1 Pneumonia

Stroke associated Pneumonia (SAP) is identified by the presence of a secondary diagnosis of hospital-acquired pneumonia and ventilator-associated pneumonia. A 2-year retrospective review of acute stroke patients found that 8.5% developed pneumonia and typically emerged within the first 4 days of hospitalization [18]. The significant risk factors of pneumonia during hospitalization include nasogastric tube, noninvasive mechanical ventilation, invasive mechanical ventilation, length of stay >2 weeks, hemorrhagic conversion, and epilepsy. A simple tool to assess the risk of pneumonia is the Integer-based Pneumonia Risk (ISAN) score is a clinical tool used to predict the likelihood of a patient developing SAP. ISAN stands for four key factors used to calculate the score: Impaired consciousness, Severe disability, Age, and Nasogastric tube use [19].

2.4.1.1 Prevention

1. Accumulation of secretions in the subglottic space, originating from the oropharynx or digestive tract, may bypass the cuff seal and enter the trachea, thereby increasing the risk of infection. Recent research indicates that subglottic secretions containing mucin can impair neutrophil activity, further compromising the host's defense mechanisms. Routine subglottal suctioning can thereby prevent their entry into the trachea and reduce the associated risk of infection. It is essential to always use a cuffed tracheostomy tube with a subglottic suction port.
2. Postural modification may aid in the prevention of aspiration in patients with dysphagia. A reclining position eases the passage of the food because of gravity. This method is part of the routine of management to prevent aspirations, which is a primary mechanism for developing pneumonia.
3. Rehabilitation programs that prioritize mobilization can significantly reduce the duration of mechanical ventilation, shorten intensive care unit (ICU) stays, decrease ventilator-associated pneumonia (VAP) rates, and improve functional recovery. These benefits are primarily attributed to the positive impact on diaphragmatic positioning, lung mechanics, and overall respiratory muscle function. An upright sitting position allows for greater diaphragmatic excursion during inhalation, leading to enhanced lung expansion. Additionally, sitting facilitates the activation of accessory respiratory muscles, such as the intercostals and upper chest muscles, further promoting lung expansion.

Prioritizing airway clearance techniques to minimize mechanical ventilation duration is crucial, as ventilator use itself is a risk factor for VAP. Respiratory physiotherapy, potentially augmented with inspiratory muscle training, may further improve inspiratory muscle function and accelerate weaning from mechanical ventilation [18].

2.4.1.2 Rehabilitation

Along with antibiotics, various strategies are employed to enhance patient outcomes. These strategies include positioning, manual hyperinflation, vibration, suctioning, and general mobilization. Some protocols have been suggested for intensive therapy with multiple sessions per day, while others use a more conservative approach with fewer sessions. Therapy intensity and frequency are typically customized based on the individual patient's needs, as determined by a physiotherapist's assessment. Manual hyperinflation is particularly beneficial in improving lung compliance, oxygenation, and collateral ventilation, especially when performed with an inspiratory pause. This technique is valuable, as it aids in preventing the collapse of terminal respiratory units when airways are obstructed. Additionally, manual hyperinflation enhances airway secretion clearance by a quick release of the bag; this technique increases expiratory flow and mimics forced expiration, facilitating the movement of secretions toward central airways. Furthermore, slow expiratory clearance techniques can be combined with manual hyperinflation. These techniques generate an expiratory flow that exceeds inspiratory flow, which is effective for mobilizing secretions from peripheral airways to central airways. Such techniques are particularly useful for addressing the lung parenchyma, where mucus build-up can obstruct airflow and exacerbate respiratory complications [20].

In some approaches, specific methods like rib springing were employed to stimulate deeper breaths and mobilize secretions, while other interventions focused on basic positioning and suctioning for effective airway management. These varied techniques in respiratory rehabilitation highlight the importance of a personalized approach to care, where interventions are adjusted to optimize respiratory function based on the patient's specific clinical presentation.

The most commonly used and low cost techniques suitable for low-resource settings are incentive spirometer, which promotes deep breathing, prevents atelectasis, and strengthens respiratory muscles; Expiratory Muscle Strength Training (EMST), which strengthens expiratory muscles, improves cough and swallowing, and reduces aspiration risk; and cough assist devices, which clear secretions by simulating a natural cough, reducing infection risk, and aiding airway management. These tools support respiratory function, prevent complications, and enhance recovery [21, 22].

2.4.2 Shoulder Subluxation

Shoulder subluxation occurs when the biomechanical factors stabilizing the glenohumeral joint are compromised, often due to muscle weakness (biceps, pectorals, latissimus, subscapularis), in individuals with hemiplegia. This leads to a change in the glenoid fossa angle, causing the humerus to slide downward and outward. Ryerson and Levit identified three types of subluxations in hemiplegic individuals: inferior, anterior, and superior. Shoulder subluxation occurs in approximately 6% of stroke survivors during inpatient rehabilitation, with a higher prevalence observed in patients with middle cerebral artery strokes [23].

2.4.2.1 Prevention

- Assessment of secondary alignment problems (rib cage, scapular, humeral position).
- For stroke survivors at risk of shoulder subluxation, firm support devices (e.g., devices such as a laptray) may be used. A sling may be used when standing or walking. A shoulder pulled-up brace may also be used.
- Prevention of shoulder capsule stretch, including support and positioning while the client sits, stands, and practices walking.
- Educating all staff, carers, and stroke survivors can help prevent the occurrence of shoulder subluxation and pain resulting from weakness [24].

2.4.2.2 Rehabilitation

Research has not yet proven any effective strategy in significantly reducing shoulder subluxation following stroke. Despite this, several approaches show potential benefits. Approaches such as electrical stimulation have been widely used, and optimal application remains unclear. Yet several systematic reviews have reported that, with either short or long daily treatment durations, particularly using neuromuscular or functional electrical stimulation, shoulder subluxation can be reduced in persons with acute and subacute stroke [25–28]. Studies suggest that kinesiology taping may be beneficial in enhancing upper limb function in stroke patients by correcting shoulder subluxation and improving general disability and upper extremity function. Longitudinal strapping has shown promise in addressing shoulder subluxation and pain. While the direct impact on functional outcomes may vary, strapping can raise awareness of the affected limb among patients and caregivers, potentially leading to indirect benefits [25].

Additionally, a suprascapular nerve block provides targeted pain relief by blocking nociceptive signals from the shoulder joint, making it an effective option when pain significantly limits participation in therapy. This minimally invasive procedure improves comfort and facilitates active rehabilitation. Clinical practice guidelines for managing such pain are straightforward and well suited for implementation in low-resource settings. They include:

- Early prevention: Proper positioning, gentle range-of-motion exercises, and shoulder support to prevent subluxation.
- Pain relief: Use of analgesics, muscle relaxants, or nerve blocks as needed.
- Therapeutic interventions: Incorporate physical therapy, neuromuscular electrical stimulation, and task-specific training to restore function.
- Adjunct therapies: Techniques like taping, slings, or acupuncture for additional support.
- Individualized approach: Regular assessment to tailor strategies based on pain severity and rehabilitation goals.
- A multidisciplinary approach ensures optimal pain control and maximized functional recovery.

2.4.3 Deep Vein Thrombosis

Stroke patients, particularly those with significant leg weakness and immobility, are at heightened risk for deep venous thrombosis (DVT) and pulmonary embolism due to increased thrombin formation and heightened platelet activity.

2.4.3.1 Prevention Strategies as Recommended by AHA/ASA Guidelines for DVT Prevention in Acute Stroke

- Pharmacological prophylaxis: Use low-molecular-weight or unfractionated heparin for immobile patients, balancing DVT prevention with bleeding risk. However pharmacological prophylaxis should not be used in the first 24 hours after thrombolysis until brain imaging has excluded significant hemorrhagic transformation [28].
- Mechanical prophylaxis: Use intermittent pneumatic compression (IPC) for patients with contraindications to anticoagulants; AVOID routine use of graduated compression stockings. For stroke patients receiving intermittent pneumatic compression, skin integrity should be assessed daily.

- Early mobilization: Early mobilization and adequate hydration should be encouraged to help prevent DVT and reduce thromboembolic risks.
- Individualized approach: Tailor prevention strategies based on mobility, bleeding risk, and overall condition [26].

2.4.4 Pressure Sores

Pressure injuries, previously known as pressure ulcers or sores, are localized areas of skin or underlying soft tissue damage caused by prolonged pressure, shear, or friction. Following acute stroke admission, approximately 3.9% of patients develop pressure injuries. Risk factors include advanced age, greater stroke severity, immobility, incontinence, poor nutritional status, and diabetes. Risk assessment tools, such as the Braden, Norton, and Waterlow scales, offer moderate sensitivity and specificity and may be more effective than clinical judgment alone in identifying stroke survivors at high risk for pressure injuries.

2.4.4.1 Prevention

- Perform a comprehensive pressure injury risk assessment based on initial screening results after admission and following any changes in the patient's condition.
- Account for additional risk factors in critically ill patients, such as length of stay in critical care, mechanical ventilation, and vasopressor use, which may increase the likelihood of pressure injuries.
- When assessing pressure injury risk: apply a structured approach, complete a thorough skin examination, complement the use of risk assessment tools with evaluation of additional risk factors, and interpret findings with clinical judgment.
- Position the patient to achieve optimal offloading of all bony prominences and to maximize pressure redistribution.

2.4.4.2 Rehabilitation

Evidence on pressure sores in stroke patients is limited; however, substantial data exists on their occurrence in other neurological and geriatric conditions, such as:

- Mobilization: if mobilization is feasible, initiate graded mobilization while ensuring that the affected area is appropriately offloaded to prevent further pressure or damage.
- Positioning: Continue to offload all bony prominences and maximize pressure redistribution by positioning at adequate intervals [26].
- Low-level laser therapy (LTT): LLT consists of amplified light of low radiation power capable of promoting biochemical, bioelectric, and bioenergetic effects resulting in stimulation of microcirculation, analgesia, anti-inflammatory and anti-edematous effects, and healing. It promotes the growth of epithelial cell motility and collagen synthesis, which contributes to the healing of pressure sores. Wavelength is an important parameter of LTT to promote healing, a wave-

length of 658 nm with a dose of 4 J/cm [2] was found to have a significant effect on the healing rate of pressure sore.
- Ultraviolet C rays (UVC-R): UVC has been proven beneficial to treat stage 2 pressure ulcers.

No evidence of benefit exists for ultrasound therapy in the treatment of pressure ulcers [29].

2.4.5 Swelling of the Extremities

Individuals who are upright (standing or sitting) with a weak, immobile arm or leg are at risk for hand or foot swelling. Evidence for effective prevention and treatment interventions is limited. However, manual edema mobilization, including massage (proximal-to-distal and distal-to-proximal), exercises, pump point stimulation, and low-stretch bandaging, has shown potential for reducing hand edema in subacute and chronic cases based on low- to moderate-quality evidence in mixed populations Conventional methods, such as limb elevation, exercise, and compression, remain primary approaches for managing swelling. For stroke survivors with severe weakness, management may include passive mobilization and limb elevation during rest, though further research is warranted to validate these strategies [23].

2.5 Recommended Outcome Measures in Acute Settings for Post-Stroke Rehabilitation in Low-Resource Settings

Outcome measurement in acute post-stroke rehabilitation is essential for assessing recovery progress, guiding treatment adjustments, and ensuring that rehabilitation strategies yield meaningful functional improvements. However, implementing robust outcome measures in low-resource settings presents unique challenges, including limited access to specialized equipment, standardized assessments, and trained personnel. This section examines recommended outcome measures for acute post-stroke rehabilitation in resource-constrained settings, highlights relevant national and regional programs, and provides educational resources tailored for low-resource environments. Furthermore, alternative approaches and point-of-care solutions for low- and middle-income countries (LMICs) are discussed to offer practical tools that can drive effective, outcome-based care (Table 2.2).

2.5.1 Recommended Outcome Measures in Low-Resource Acute Settings

Acute stroke rehabilitation requires outcome measures that accurately capture patient progress across multiple domains, including motor function, cognitive recovery, and quality of life. Ideal tools are standardized, quick to administer, and

Table 2.2 Recommended outcome measures in post-stroke acute phase in low-resource settings

International classification of impairment, disability, and handicap: domains	Outcome measure recommended
Body structure and function	Fugl-Meyer assessment (FMA)—Arm and leg
Activity	Action research arm test (ARAT) and 10-meter walk test
Quality of life	NA
Global disability	NA
Participation	NA
Others	Trunk control test (TCT), trunk impaired scale (TIS)
Fall risk	BBS or Morse scale
Mental state	Montreal cognitive assessment if available and suitable in the local context

adaptable to the limitations of resource-constrained environments. Commonly used outcome measures, such as the modified Rankin Scale (mRS) and modified Barthel Index, are widely recognized for their ability to evaluate functional independence, motor recovery, and neurological deficits. These tools are validated and have been adapted successfully in low-resource settings.

1. National Institutes of Health Stroke Scale (NIHSS).
   The NIHSS evaluates neurological function in domains like consciousness, motor strength, and language. Although typically applied in high-resource settings, it has been successfully adapted for LMICs with simplified training for community health workers. Some stroke centers in LMICs have modified the NIHSS to focus on the most relevant aspects of motor and cognitive recovery, ensuring feasibility without sacrificing reliability. Examples include adaptations in parts of India, where stroke registries have applied this measure with the help of mobile applications to streamline data collection [30].Although recommended for measuring stroke severity, it should not be used as a measure of recovery.
2. Modified Rankin Scale (mRS).
   The mRS is a simple, reliable measure of functional independence post-stroke. This ordinal scale ranges from 0 (no symptoms) to 6 (death), allowing healthcare providers to classify the level of disability. In low-resource settings, the mRS is advantageous because it requires minimal equipment and can be administered quickly by a trained professional. In settings where professionals are scarce, community health workers can be trained to apply this measure reliably, as seen in rural stroke programs in South America and Southeast Asia [31].
3. Modified Barthel Index (mBI).
   The Barthel Index assesses a patient's ability to perform activities of daily living (ADLs), providing insight into the patient's functional capacity. It scores ten areas of ADLs using a 5-point rating scale rather than the original 3-point rating scale. Additionally, the internal consistency reliability coefficient for the modified scoring of the Barthel Index was 0.90, compared to 0.87 for the original scale [32].

4. 10-Meter Walk Test (10 MWT)
   The 10-MWT is a simple and reliable measure of walking speed, which is a strong predictor of functional independence and community ambulation post-stroke. The test requires minimal equipment: a marked 10-m walkway and a stopwatch. Patients are instructed to walk at a comfortable or maximum safe speed, and the time taken to cover the central 6 m is recorded to eliminate acceleration and deceleration effects. The tool has been validated across the acute, subacute, and chronic phases of stroke [33].
5. Fugl-Meyer Assessment (FMA).
   The FMA was designed by Fugl-Meyer et al. to provide a numeric score of motor status after stroke, based on the sequential stages of motor recovery and Brunnstrom, using measures such as limb synergies and range of motion. It is both valid and reliable across multiple studies. The scale has a maximum score of 226 points, with the upper limb (66 points) and lower limb (34 points) subsections most frequently used in rehabilitation programs [34].

The International Stroke Recovery and Rehabilitation Alliance (ISRRA) has developed consensus-based core recommendations for ICIDH-aligned assessments in the acute post-stroke phase [35]. These include the Fugl-Meyer Assessment for evaluating body structure and function of the arm and leg, the Action Research Arm Test for assessing activity limitations of the arm, and the 10-Meter Walk Test for evaluating walking ability. Domains such as quality of life, global disability, and participation are deemed unnecessary for evaluation during the acute phase.

2.5.2 Alternatives to Ideal Outcome Measures in LMICs

In environments where standardized tools are unavailable or infeasible, alternative point-of-care solutions provide practical substitutes. For instance, if the NIHSS cannot be fully implemented, a pared-down checklist focusing on critical items like motor function and language ability can serve as a proxy assessment. Community health workers can be trained in these simplified protocols, improving consistency in outcome measurement without requiring extensive resources [36].

Mobile-based assessments are also emerging as viable alternatives for outcome measurement. Mobile applications that use motion sensors to evaluate motor function or text-based prompts for ADL assessments can enable consistent outcome monitoring. Some LMIC programs are leveraging these mobile technologies to conduct regular follow-ups with patients remotely, reducing travel burdens and increasing follow-up rates [36].

The Trunk Control Test (TCT) is a validated tool for assessing trunk balance in patients with significant impairments. Each item is scored as 0 (unable to perform without assistance), 12 (able to perform with an abnormal style), or 25 (able to perform normally). For the sitting balance item, a score of 12 is assigned if hand support is required, and 0 if the patient cannot remain upright for 30 s. The TCT at admission has demonstrated predictive value for discharge outcomes and serves as a reliable functional predictor. Its application in low- and middle-income countries (LMICs) can support the development of customized rehabilitation protocols.

2.6 National and Regional Programs/Networks for Stroke Rehabilitation

Several LMICs have developed programs that provide scalable acute rehabilitation services, incorporating feasible outcome measures to support their unique environments.

1. *SARAH Network, Brazil.*
   The SARAH Network in Brazil is a notable example, delivering integrated rehabilitation services for stroke survivors in acute and subacute phases. SARAH incorporates cost-effective assessment tools, such as the mRS and Barthel Index, along with culturally adapted outcome measures for psychological and cognitive assessments. By training a multidisciplinary team to apply these assessments, SARAH successfully evaluates and tracks patient outcomes, demonstrating that structured yet adaptable outcome measures can thrive even in resource-constrained settings [37].
2. *The National Stroke Program, Ghana.*
   Ghana's National Stroke Program utilizes community health workers to support outcome measurement and data collection in acute post-stroke rehabilitation. Community health workers are trained in the Barthel Index and a simplified NIHSS, which they administer during routine follow-up visits. This decentralized approach improves access to rehabilitation for patients in rural areas while maintaining consistent monitoring of recovery progress [38].
3. *The Indian Stroke Rehabilitation Registry.*
   In India, the Stroke Rehabilitation Registry was established to collect and analyze data on post-stroke recovery and outcomes. Utilizing the Barthel Index and NIHSS, the registry supports acute stroke care in remote regions through mobile health technology, where health workers use tablets to record patient progress. This model has enabled large-scale data collection, contributing valuable insights into stroke outcomes in low-resource settings [39].
4. *The Indian Stroke Clinical Trial Network (INSTRuCT).*
   The Indian Stroke Clinical Trial Network (INSTRuCT), comprising 59 hospitals across India, facilitates clinical trials and research in stroke treatment, prevention, and rehabilitation. Established under the Indian Council of Medical Research (ICMR), it also provides training for stroke physicians and trial coordinators. Key trials include ENCHANTED, an international study that found no significant benefit of intensive blood pressure lowering (<140 mm Hg) over standard care (<180 mm Hg) in acute ischemic stroke. SPRINT INDIA is assessing a structured stroke prevention package to reduce recurrent strokes and associated risks in subacute patients [40].

5. *Global Consortium of Stroke Rehabilitation (GCSR)—serving low-resource settings*: The Global Consortium of Stroke Rehabilitation (GCSR) is a collaborative initiative dedicated to advancing stroke rehabilitation in low- and middle-income countries (LMICs). The consortium brings together like-minded individuals with a shared interest in advancing stroke rehabilitation efforts from various LMICs. The consortium is currently focused on advocacy and research aimed at improving stroke rehabilitation outcomes in these regions [41].

2.7 Educational Resources for Acute Rehabilitation in Low-Resource Settings

Educational resources play an essential role in empowering both patients and healthcare providers to engage in effective rehabilitation practices. Given the scarcity of specialists in LMICs, accessible educational tools are critical.

1. *Patient education.*
   Visual aids, booklets, and mobile applications that provide step-by-step guides on post-stroke exercises and self-care routines can support patients in self-directed rehabilitation. For example, the WHO's Rehabilitation 2030 initiative provides simple, illustrated guides for stroke rehabilitation exercises tailored to resource-constrained settings [42].
2. *Healthcare professional training.*
   In-service training programs focused on acute rehabilitation have been developed to address skill gaps among healthcare providers in LMICs. Online courses and video modules, such as those offered by the SARAH Network, train providers in administering outcome measures and conducting basic rehabilitation exercises. These resources support healthcare workers in delivering quality care with limited resources, fostering skill development across various professional roles [37].

2.8 Conclusion

The medical aspects of acute post-stroke rehabilitation require a multidisciplinary approach, with neurologists and physiatrists playing crucial roles in stabilizing patients, preventing complications, and initiating early recovery interventions. In LMICs, where resources are limited, innovative strategies such as task-shifting, telemedicine, and standardized care pathways can bridge gaps in service delivery. By tailoring evidence-based interventions to local contexts, it is possible to improve functional outcomes and reduce the burden of stroke in low-resource settings.

What Does Evidence Say

- Stroke rehabilitation should be structured to deliver the maximum feasible amount of scheduled therapy, including physiotherapy and occupational therapy.
- Patients should begin mobilizing out of bed within 48 h of stroke onset unless contraindicated, but intensive out-of-bed activity should be avoided within the first 24 h.
- Skin assessments using a validated risk scale (e.g., Braden Scale) should be completed within 8 h of admission and repeated after any clinical change.
- Patients with a high risk of pressure injuries should receive preventive care, including repositioning every 4–6 h, barrier creams, and pressure-relieving devices.
- Prophylactic-dose subcutaneous heparin should be used in ischemic stroke patients during the acute and rehabilitation phase or until mobility improves.
- Antithrombotic stockings are not recommended for the prevention of DVT or pulmonary embolism post-stroke.
- Swallow screening should be completed as early as possible and within 24 h of admission.
- All patients should be screened for malnutrition at admission and monitored regularly (at least weekly) during hospitalization.
- Suspected urinary incontinence should be evaluated using a structured assessment by trained personnel.
- If present, a documented continence management plan should be implemented; routine use of indwelling catheters is discouraged.
- Suspected fecal incontinence or constipation should be assessed systematically.
- Identified issues should be managed with a structured, documented bowel care plan.
- A clear discharge plan should be documented, including ongoing therapy needs, care arrangements, and follow-up plans.

References

1. Bernhardt J, Hayward KS, Kwakkel G, Ward NS, Wolf SL, Borschmann K, Krakauer JW, Boyd LA, Carmichael ST, Corbett D, Cramer SC. Agreed definitions and a shared vision for new standards in stroke recovery research: the stroke recovery and rehabilitation roundtable taskforce. Int J Stroke. 2017;12(5):444–50. https://doi.org/10.1177/1747493017711816. PMID: 28697708.
2. Surya N, Srivastava A. Stroke rehabilitation. In: Narsimhan L, editor. Reviews in neurology 2019 stroke. London: IAN Publications; 2019. p. 295–307.

3. Liu R, Liu J. Prognostic factors of functional outcome in post-acute stroke in the rehabilitation unit. J Formos Med Assoc. 2022;121(2):568–9. https://doi.org/10.1016/j.jfma.2021.08.008. Epub 2021 Aug 20.
4. Stinear CM, Byblow WD, Ackerley SJ, Smith MC, Borges VM, Barber PA. PREP2: a biomarker-based algorithm for predicting upper limb function after stroke. Ann Clin Transl Neurol. 2017;4(11):811–20. https://doi.org/10.1002/acn3.488. PMID: 29159193; PMCID: PMC5682112.
5. Smith MC, Barber PA, Stinear CM. The TWIST algorithm predicts time to walking independently after stroke. Neurorehabil Neural Repair. 2017;31(10-11):955–64. https://doi.org/10.1177/1545968317736820. Epub 2017 Nov 1. PMID: 29090654.
6. Langhorne P, et al. Stroke rehabilitation in low-resource settings. Geneva: World Health Organization; 2020.
7. Feigin VL, et al. Global burden of stroke and risk factors. Lancet Neurol. 2018;17(11):877–94.
8. Anderson CS, et al. Intensive blood pressure reduction in acute cerebral hemorrhage trial. Lancet. 2017;385(9976):1454–63.
9. Bernhardt J, et al. Early rehabilitation after stroke. Stroke. 2015;46(8):2185–94.
10. Langhorne P, Wu O, Rodgers H, Ashburn A, Bernhardt J. A very early rehabilitation trial after stroke (AVERT): a phase III, multicentre, randomised controlled trial. Health Technol Assess. 2017;21(54):1–120. https://doi.org/10.3310/hta21540. PMID: 28967376; PMCID: PMC5641820.
11. PLOS ONE. Nurse-led telerehabilitation intervention to improve stroke efficacy. PLoS One. 2021;18(6):e0280973.
12. Surya N, Someshwar HP. Low-cost telerehabilitation in low- and middle-income countries (LMICs): overcoming barriers to access and improving healthcare delivery. NeuroRehabilitation. 2025;56(1):30–6. https://doi.org/10.1177/10538135241303349.
13. Jones-Esan L, Somasiri N, Lorne K. Enhancing healthcare delivery through digital health interventions: a systematic review on telemedicine and mobile health applications in low and middle-income countries (LMICs). Res Sq. 2024; https://doi.org/10.21203/rs.3.rs-5189203/v1.
14. Labeit B, Michou E, Hamdy S, Trapl-Grundschober M, Suntrup-Krueger S, Muhle P, Bath PM, Dziewas R. The assessment of dysphagia after stroke: state of the art and future directions. Lancet Neurol. 2023;22(9):858–70. https://doi.org/10.1016/S1474-4422(23)00153-9. PMID: 37596008.
15. Chang M, Wang Q, Jiang Y, Shi H, Peng T, Wang M. Optimization of early mobilization program for patients with acute ischemic stroke: an orthogonal design. Front Neurol. 2021;12:645811. https://doi.org/10.3389/fneur.2021.645811. PMID: 33912126; PMCID: PMC8072335.
16. National Institute for Health and Care Excellence (NICE). Stroke rehabilitation in adults: movement. London: NICE; 2023. Available from: https://www.nice.org.uk/guidance/ng236/chapter/Recommendations#movement.
17. Wang PL, Zhao XQ, Yang ZH, et al. Effect of in-hospital medical complications on case fatality post-acute ischemic stroke: data from the China National Stroke Registry. Chin Med J. 2012;125(14):2449–54.
18. Patel UK, Kodumuri N, Dave M, et al. Stroke-associated pneumonia: a retrospective study of risk factors and outcomes. Neurologist. 2020;25(3):39–48. https://doi.org/10.1097/NRL.0000000000000269.
19. Akimoto T, Hara M, Ishihara M, Ogawa K, Nakajima H. Post-stroke pneumonia in real-world practice: background, microbiological examination, and treatment. Neurol Int. 2023;15(1):69–77. https://doi.org/10.3390/neurolint15010006.
20. Smith CJ, Bray BD, Hoffman A, Meisel A, Heuschmann PU, Wolfe CD, Tyrrell PJ, Rudd AG. Can a novel clinical risk score improve pneumonia prediction in acute stroke care? A UK multicenter cohort study. J Am Heart Assoc. 2015; https://doi.org/10.1161/JAHA.114.001307.

21. Coppadoro A, Bellani G, Foti G. Non-pharmacological interventions to prevent ventilator-associated pneumonia: a literature review. Respir Care. 2019;64(12):1586–95. https://doi.org/10.4187/respcare.07127.
22. Pozuelo-Carrascosa DP, Torres-Costoso A, Alvarez-Bueno C, Cavero-Redondo I, López Muñoz P, Martínez-Vizcaíno V. Multimodality respiratory physiotherapy reduces mortality but may not prevent ventilator-associated pneumonia or reduce length of stay in the intensive care unit: a systematic review. J Physiother. 2018;64(4):222–8.
23. Lee JH, Baker LL, Johnson RE, Tilson JK. Effectiveness of neuromuscular electrical stimulation for management of shoulder subluxation post-stroke: a systematic review with meta-analysis. Clin Rehabil. 2017;31(11):1431–44.
24. Ada L, Foongchomcheay A. Efficacy of electrical stimulation in preventing or reducing subluxation of the shoulder after stroke: a meta-analysis. Aust J Physiother. 2002;48(4):257–67. https://doi.org/10.1016/s0004-9514(14)60165-3.
25. Wang Y, Li X, Sun C, Xu R. Effectiveness of kinesiology taping on the functions of upper limbs in patients with stroke: a meta-analysis of randomized trial. Neurol Sci. 2022;43(7):4145–56. https://doi.org/10.1007/s10072-022-06010-1.
26. Australian and New Zealand living clinical guidelines for stroke management.
27. Gandhi DB, Sterba A, Khatter H, Pandian JD, Bhanot K. Mirror therapy in stroke rehabilitation: why, how early, and effects: a meta-analysis. J Stroke Med. 2020;3(2):72–80. https://doi.org/10.1177/2516608520974512.
28. Powers WJ, Rabinstein AA, Ackerson T, Adeoye OM, Bambakidis NC, Becker K, Biller J, Brown M, Demaerschalk BM, Hoh B, Jauch EC, Kidwell CS, Leslie-Mazwi TM, Ovbiagele B, Scott PA, Sheth KN, Southerland AM, Summers DV, Tirschwell DL. Guidelines for the early management of patients with acute ischemic stroke: 2019 update to the 2018 guidelines for the early management of acute ischemic stroke: a guideline for healthcare professionals from the American Heart Association/American Stroke Association. Stroke. 2019;50(12):e344–418. https://doi.org/10.1161/STR.0000000000000211.
29. Machado RS, Viana S, Sbruzzi G. Low-level laser therapy in the treatment of pressure ulcers: systematic review. Lasers Med Sci. 2017;32(4):937–44. https://doi.org/10.1007/s10103-017-2150-9.
30. Saxena S, et al. Adaptations of NIHSS for community healthcare in India. Indian J Neurol. 2020;27(2):145–53.
31. Guzik A, et al. Modified Rankin scale application by non-specialists: a rural pilot. J Neurol Res. 2021;28(4):125–30.
32. Nijensohn D, et al. Reliability and application of Barthel index in resource-limited settings. Rehabil Sci. 2018;5(3):215–22.
33. Busk H, Holm P, Skou ST, Seitner S, Siemsen T, Wienecke T. Inter-rater reliability and agreement of 6 minute walk test and 10 meter walk test at comfortable walk speed in patients with acute stroke. Physiother Theory Pract. 2023;39(5):1024–32.
34. Duncan PW, Propst M, Nelson SG. Reliability of the Fugl-Meyer assessment of sensorimotor recovery following cerebrovascular accident. Phys Ther. 1983;63(10):1606–10.
35. Kwakkel G, Lannin NA, Borschmann K, et al. Standardized measurement of sensorimotor recovery in stroke trials: consensus-based core recommendations from the stroke recovery and rehabilitation roundtable. Int J Stroke. 2017;12(5):451–61. https://doi.org/10.1177/1747493017711813.
36. Broussalis E, et al. Assessing functional outcomes in stroke patients: applicability of simplified scales in low-resource settings. Int J Stroke. 2019;14(6):540–7.
37. Rodrigues C, et al. SARAH network: a model of low-resource stroke rehabilitation in Brazil. Braz J Neurol. 2018;35(1):45–54.
38. Adu-Boakye Y, et al. National stroke programs in Africa: outcomes and feasibility in Ghana. Afr J Public Health. 2021;12(5):310–9.

39. Patel R, et al. India's stroke rehabilitation registry: advancing outcomes with mobile health. Asia Pacific J Health Manage. 2019;14(4):212–20.
40. Anderson CS, Huang Y, Lindley RI, Chen X, Arima H, Chen G, Li Q, Billot L, Delcourt C, Bath PM, Broderick JP. Intensive blood pressure reduction with intravenous thrombolysis therapy for acute ischaemic stroke (ENCHANTED): an international, randomised, open-label, blinded-endpoint, phase 3 trial. Lancet. 2019;393(10174):877–88.
41. Gandhi DB, Kamalakannan S, Urimubenshi G, Sebastian IA, Montanaro VV, Chawla NS, Dsouza JV, Ngeh E, Mahmood A, Demers M, Hombali A. Stroke rehabilitation clinical practice guidelines in low and middle income countries-a systematic review of quality unique features. Cerebrovasc Dis. 2024;54(4):429–40.
42. World Health Organization. Rehabilitation 2030: a call for action on stroke care in low-resource settings. Geneva: WHO; 2021.

Outline: Contextualized Arm and Hand Rehabilitation

3

G. Alexander, Rinita Mascarenhas, Gerad Urimubenshi, Dorcas B. C. Gandhi, and John M. Solomon

Abstract

Approximately 80% of stroke survivors experience upper limb impairments, yet fewer than 15% regain full bilateral coordination. Recovery is often slow and incomplete, attributed to the complexity of upper limb control and the limited therapy time allocated in conventional rehabilitation. Evidence supports a range of interventions, including constraint-induced movement therapy, mirror therapy, motor imagery, progressive resistance training, repetitive task-specific training, and electrical stimulation, that can facilitate functional recovery. However, their feasibility in low- and middle-income countries (LMICs) is constrained by limited infrastructure, a lack of skilled personnel, and costly technologies. Contextualized approaches are therefore essential. Integrating culturally relevant, task-oriented training using locally available resources, caregiver-supported home practice, and community-based rehabilitation models can help bridge these

G. Alexander · J. M. Solomon (✉)
Department of Physiotherapy, Manipal College of Health Professions, Manipal Academy of Higher Education, Manipal, India
e-mail: john.solomon@manipal.edu

R. Mascarenhas
Department of Neurology, Christian Medical College and Hospital, Ludhiana, Punjab, India

G. Urimubenshi
Department of Physiotherapy, College of Medicine and Health Sciences, University of Rwanda, Kigali, Rwanda

D. B. C. Gandhi
Department of Neurology, Christian Medical College & Hospital Ludhiana, Ludhiana, Punjab, India

Department of Physiotherapy, Manipal Academy of Higher Education, Manipal, Karnataka, India

D. B. C. Gandhi, J. D Pandian (eds.), *Stroke Rehabilitation in Low-Resource Settings*, https://doi.org/10.1007/978-981-92-0403-8_3

gaps. Bilateral arm functional training, particularly when applied through structured yet low-cost strategies such as the "table wrap" approach and household-based adaptations (e.g., cooking, clothing, or household chores), promotes neuroplasticity, overcomes learned nonuse, and supports participation in daily life. Caregiver training and balanced involvement are vital to sustain motivation and autonomy. This chapter highlights the importance of evidence-based yet context-sensitive interventions to improve upper limb recovery, independence, and quality of life in stroke survivors in resource-limited settings.

Keywords

Upper limb impairment • Bilateral arm training • Task-oriented practice • Low-resource settings

Learning Objectives

1. To understand the impairments and complications of the upper limb following hemiplegia due to stroke.
2. To learn evidence-based strategies to improve upper limb recovery.
3. To understand the challenges in the implementation of upper limb programs in low-resource settings and explore possible solutions.
4. To explore models of contextualized programs and recent novel approaches that could be contextualized for implementation in low-resource settings.

3.1 Introduction

Stroke significantly impacts the quality of life, often due to the significant disability caused by hemiplegia [1]. This leads to a significant burden and disability in low- and middle-income countries compared to high-income countries due to limited awareness, lack of resources, inadequate rehabilitation, and insufficient support from the community and government.

Motor rehabilitation focuses on alleviating this impairment and is defined as "a process that engages people with stroke in order to benefit their motor function, activity capacity and performance in daily life. It is necessary for all people with residual motor disability whose goal is to enhance their functioning, independence and participation" [2]. One of the most challenging aspects of motor rehabilitation achieving recovery and functional use of the impaired upper limb after a stroke. Unlike the lower limb, upper limb recovery is often slow and incomplete. This is often attributed to the complexity of upper limb movements that require a greater amount of training compared to the lower limb. Most often, therapy time is spent on achieving mobility during the initial hospital stay, with very little time spent on upper limb therapy. Lang et al.'s study demonstrated that even in a well-established rehabilitation center, the amount of repetitions achieved by therapists for the upper limb is only around 30–40, which is far less than the required 300 repetitions per day.

Recent evidence has demonstrated that a structured, intensive upper limb training program that achieves numbers close to this can bring about recovery even in the chronic stage. Some of these programs are developed and implemented through an interdisciplinary team in well-equipped setups. In low-resource settings, implementing such a program still remains a challenge.

Understanding the recovery process and modifying evidence-based programs in a contextualized manner could bring about better outcomes in stroke survivors with upper limb impairment.

3.2 Post-Stroke Upper Limb Impairments

Approximately 80% of individuals who experience a stroke suffer from upper limb impairments, yet only 5% achieve full recovery [3]. The recovery process begins immediately post-stroke, following spontaneous neurological recovery, and typically plateaus 3–6 months after the stroke [4]. Upper limb function can improve through either true recovery or compensatory mechanisms. True recovery involves the reactivation of neuronal structures in the injured core, penumbra, or diaschisis regions at the cellular level. At the activity performance level, it reflects a reduction in impairment and a return to normal movement patterns present before stroke. At the participation level, it involves completing upper limb tasks similarly to healthy individuals. Compensation, in contrast, involves the activation of alternative neuronal areas not typically utilized in healthy individuals. At the activity level, this manifests as alternative movement strategies, including altered joint play or muscle activation to complete tasks in a novel way. At the participation level, it often involves completing tasks using the non-paretic limb or other body parts, rather than engaging in bimanual activities [5].

Motor recovery after stroke is nonlinear, reflecting the complex relationship between brain organization and behavioral improvement through therapeutic interventions. The impairments are not static; as motor recovery progresses, the nature of impairment changes and treatment needs vary over time. Additionally, multiple impairments, such as pain, increased muscle tone, muscle weakness, and contractures, may develop. If left untreated, these can negatively impact the recovery process. Furthermore, there is often a mismatch between objectively observed upper limb function and patients' subjective perceptions of their abilities following stroke. Stroke survivors frequently overestimate or underestimate their functional capabilities, which can influence their participation in rehabilitation and daily activities. This mismatch may result from factors such as cognitive impairments, depression, or altered body awareness commonly seen in stroke survivors.

Upper limb impairment following a stroke significantly disrupts daily life functions, primarily due to a lack of bilateral coordination. Evidence report that the recovery rate for such impairments is less than 15%, highlighting the complexity of restoring bilateral arm coordination. This challenge is particularly pronounced in low-resource settings, where access to advanced rehabilitation is limited. In these regions, bilateral arm functional training could have a profound impact on stroke survivors, as the ultimate goal of rehabilitation is to enable the effective use of both arms for daily activities.

Three functional behaviors are often observed post-stroke: learned nonuse, learned bad use, and forgetting behavior. Learned nonuse occurs when individuals progressively stop using their affected upper limb despite having the potential to move it. Initially, nonuse arises due to impairments such as weakness, paralysis, or sensory loss. Over time, this lack of usage becomes habitual, and the affected limb is excluded from functional activities, evolving into a learned pattern that further limits recovery. Learned bad use happens when individuals adopt compensatory strategies to perform tasks with the paretic limb. Impairments such as weakness, sensory deficits, pain, stiffness, spasticity, and abnormal motor synergies contribute to these compensations. Common examples include using trunk flexion instead of elbow extension, or wrist flexion instead of wrist extension for reaching and grasping tasks. While these strategies may initially succeed, they often lead to poor accuracy and reinforce abnormal movement patterns, reducing performance despite continued training. Forgetting behavior refers to the decline of motor skills acquired through training during periods of no practice. Unlike in healthy individuals, stroke survivors experience more transient retention of motor skills, necessitating additional training to regain proficiency [6].

Upper limb impairment significantly disrupts daily life by limiting bilateral coordination. Cauraugh et al. reported that the recovery rate for bilateral coordination after stroke is less than 15%, underscoring the complexity of restoring functional use of both arms. This challenge is even more pronounced in low-resource settings, where access to advanced rehabilitation technologies is limited. In such contexts, bilateral arm functional training offers a promising and practical approach, as the ultimate goal of stroke rehabilitation is to enable the effective use of both arms for meaningful daily activities.

Sensory impairments in the upper limb, such as numbness, altered touch perception, and pain, significantly disrupt daily activities and the ability to perform tasks requiring fine motor control. Studies report that sensory impairments are not adequately understood by health professionals or social circles, and there is a lack of specific rehabilitation strategies targeting sensory impairments [7].

3.3 Secondary Complications and their Prevention

Secondary complications in the upper extremity (UE) are common in patients with hemiplegia and pose significant challenges to rehabilitation. These complications often stem from biomechanical and neurological factors, which can exacerbate pain and functional limitations, impacting recovery. Among these, shoulder subluxation and complex regional pain syndrome (CRPS) are particularly prevalent and disabling. A comprehensive understanding of these complications, along with targeted prevention strategies, is essential for optimizing rehabilitation outcomes.

3.3.1 Shoulder Subluxation

Shoulder subluxation is one of the most common complications affecting the hemiplegic upper extremity. It occurs when the biomechanical factors maintaining glenohumeral joint stability are disrupted. In hemiplegic patients, muscle weakness in the shoulder girdle alters the angle of the glenoid fossa, which is normally held at 40° in the frontal plane. This biomechanical imbalance destabilizes the shoulder joint, leading to subluxation. Without proper management, subluxation increases the risk of microtrauma and soft tissue injuries, further complicating the rehabilitation process.

3.3.2 Complex Regional Pain Syndrome (CRPS)

Complex regional pain syndrome, also known as shoulder–hand syndrome, is one of the most disabling secondary complications. It affects over 12.5% of hemiplegic patients and significantly hampers rehabilitation efforts. CRPS is characterized by severe shoulder pain, hand swelling, vasomotor and sudomotor instability, joint contractures, and functional limitations of the affected extremity. While the precise cause remains unclear, risk factors include biomechanical instability, shoulder subluxation, and repeated microtraumas, which may trigger chronic pain and the development of an abnormal sensory-sympathetic reflex arc or dorsal horn neuron sensitization.

In addition to shoulder subluxation and CRPS, hemiplegic patients are susceptible to complications such as adhesive capsulitis and spasticity. Each of these conditions can further impair functional recovery and complicate rehabilitation programs (refer to Chap. 8).

3.3.3 Prevention Strategies

Preventing complications in the upper extremity requires a proactive and multifaceted approach. Early initiation of exercises targeting the shoulder stabilizers is essential. Proper positioning plays a critical role in avoiding microtrauma and shoulder capsule stretching; the affected shoulder should be supported during sitting, standing, and walking with the use of slings, lap trays, or pillows. However, hand splints are not recommended; such measures help maintain joint integrity and reduce the risk of further injury during daily activities and ambulation. Although the evidence for the use of electrical stimulation and taping remains limited, these modalities may be considered as adjuncts, as their use is unlikely to cause harm.

3.4 Assessment of Upper Limb Deficits Post-Stroke

Any assessment is incomplete without the use of the International Classification of Functioning, Disability and Health (ICF) developed by the World Health Organization. The purpose of using the ICF framework for upper limb assessment post-stroke is to enable an understanding of the individual's functional status, capturing not only the physical impairments of the affected limb (e.g., weakness, spasticity, loss of coordination), but also the impact on daily activities (such as reaching, grasping, or self-care tasks) and participation in social and occupational roles. Table 3.1 presents the mapping of ICF domains and categories relevant to upper limb assessment post-stroke.

The participation domain is not specific to upper limb function; it remains an essential component of post-stroke assessment. It provides insight into how upper limb impairments and activity limitations influence broader aspects of daily life, independence, and social reintegration. Furthermore, according to the Stroke

Table 3.1 ICF-based assessment framework for upper limb function post-stroke

ICF domain	Component assessed	Assessment tools/measures
Body functions and structures	Motor function (muscle strength, tone, coordination, joint mobility, muscle control)	– Fugl–Meyer assessment (FMA) (upper extremity section) – Modified Ashworth scale (MAS) for tone – Medical Research Council (MRC) scale for strength – Tardieu scale (spasticity)
	Sensory function	– Nottingham sensory assessment (NSA) – Rivermead assessment of somatosensory performance (RASP)
	Pain and edema	– VAS/NRS for pain – Circumferential measurement for edema
Activities	Functional use of the upper limb	– Action research arm test (ARAT) – Wolf Motor function test (WMFT) – Box and block test (BBT) – Nine-hole peg test (NHPT) – Motor activity log (MAL) – Chedoke arm and hand activity inventory (CAHAI)
	Activities of daily living (ADLs)	– Barthel index – Functional Independence measure (FIM) – ABILHAND questionnaire
Participation	Social and occupational participation	– Stroke impact scale (SIS) – Participation subscale – Reintegration to Normal living index (RNLI)
	Perceived use and satisfaction	– Motor activity log (amount of use and quality of movement) – Patient-specific functional scale (PSFS) – Canadian occupational performance measure (COPM)

Recovery and Rehabilitation Roundtable, both FMA and ARAT are recommended for use consistently across all stages of stroke recovery.

The use of sensors and accelerometers would provide information on movement assessment. However, a systematic review of 32 articles found that sensor-based measures show a strong need for standardization and consensus regarding clinical assessments as well as the use of specific algorithms and metrics for unsupervised measurements during daily life.

3.5 Evidence-Based Strategies for Upper Limb Post-Stroke

3.5.1 Recommendations for Upper Limb Rehabilitation after Stroke

Rehabilitation of the upper limb following stroke is essential for restoring motor function and enhancing quality of life. The selection of therapy should be guided by the severity of impairment, with evidence-based interventions facilitating recovery and improving functional outcomes (Table 3.2).

(a) Constraint-induced movement therapy (CIMT) is recommended for individuals with some active wrist and finger extension. It involves restraining the unaffected limb for at least 6 h daily while engaging the affected arm in a minimum of 2 h of active therapy per day over 2 weeks. Despite its efficacy, adherence is often low due to discomfort, interference with independence, and patient preference for less intensive protocols. Modified CIMT (mCIMT) addresses these barriers by combining structured 1-hour functional practice sessions with restraint of the less affected limb for 5 h per day, 5 days a week, over 10 weeks, making it feasible for outpatient settings and reimbursable within many care plans.

(b) Mirror therapy (MT) produces small but meaningful improvements in arm function and activities of daily living, particularly in the acute phase. It activates the mirror neuron system and ipsilateral motor pathways, enhancing corticospinal excitability and promoting neuroplasticity. MT can be applied even in flaccid limbs, with sessions ranging from 20 to 90 min based on patient progress. Some patients may struggle to focus on the mirror, and caregivers may require guidance to safely implement home-based mirror boxes. MT is cost-effective and accessible.

(c) Mental practice, or motor imagery, involves cognitively simulating movements without physical execution to stabilize or enhance motor function. Patients are assessed for suitability, taught individualized imagery techniques, and gradually integrated into active therapy. Evidence shows moderate improvements in arm activity, impairment, and daily function, particularly within the first 3 months post-stroke and in patients with some voluntary movement. Action observation shows similar benefits. Effects in patients with very severe or very mild weakness remain unclear.

Table 3.2 Upper limb stroke rehab: interventions by severity (per Australian Living Guidelines)

Severity/FMA range	Recommended interventions	Evidence strength
Severe (very limited or no voluntary wrist/finger extension)	• Robotics/mechanically assisted training • Electrical stimulation (ES) + motor training • Virtual reality/interactive gaming • Sensory-specific training (if sensory loss present)	Weak recommendation Weak recommendation Weak recommendation Weak recommendation
Moderate (some proximal/limited distal movement; partial hand use)	• Repetitive task-specific training • Mental practice + active motor training • Mirror therapy (as adjunct) • Progressive resistance training (PRT) • Assistive technology + repetitive practice	Weak recommendation Weak recommendation Weak recommendation Strong recommendation Weak recommendation
Mild (FMA ≥32–34; some active wrist and finger extension)	• Constraint-induced movement therapy (CIMT) • Repetitive task-specific training • Progressive resistance training (PRT)	Strong recommendation Weak recommendation Strong recommendation
Not routinely recommended (any severity)	• Hand/wrist orthoses (splints) • Brain stimulation (tDCS/rTMS)	Strong recommendation against Weak recommendation against

Weak recommendation—Benefits outweigh harms for the majority, but not for everyone. The majority of patients would likely want this option. Strong recommendation—Benefits outweigh harms for almost everyone. All or nearly all informed patients would likely want this option. Strong recommendation against—All or nearly all would likely decline the intervention

(d) Progressive resistance training (PRT) enhances muscle strength in stroke survivors. Systematic reviews show that progressively increasing loads (8–12 repetition maximum, ≥2 sets) produce large improvements in strength (SMD 0.98) compared to no intervention or placebo. Evidence for gains in functional activity or daily living is less clear, though strength improvements support rehabilitation.

(e) Wrist and hand orthoses do not improve upper limb function, range of motion, or reduce pain. Evidence from randomized trials (N = 126) shows negligible clinical effects, and routine use, particularly within the first 6 months post-stroke, is not recommended. Evidence for use later post-stroke is limited.

(f) Repetitive task-specific training involves repeated practice of functional tasks to improve arm and hand control. A Cochrane review found small but significant improvements in upper limb function lasting up to 6 months post-intervention.
(g) Electrical stimulation, combined with motor training, produces moderate-to-large improvements in arm activity and function compared to motor training alone or no therapy. EMG-triggered, cyclic, and sensory stimulation appear similarly effective, while passive stimulation is mainly used for preventing or managing shoulder subluxation and pain.
(h) Robotics and virtual rehabilitation offer intensive, repetitive, task-specific practice with game-based incentives that enhance motivation and engagement. Robotics allows precise control of task complexity and error augmentation to promote neuroplasticity. Evidence shows therapy durations over 2 h per day with these technologies lead to meaningful motor improvements. Mechanically assisted arm training modestly improves arm function and daily activities, particularly when delivered early post-stroke. Virtual reality and interactive gaming can enhance function when added to usual care, although their effects are similar to those of conventional therapy when matched for dose.
(i) Noninvasive brain stimulation, including transcranial direct current stimulation (tDCS) and repetitive transcranial magnetic stimulation (rTMS), does not consistently improve arm activity. Modest benefits may occur in measures of impairment or activities of daily living, particularly in chronic stroke or mild/moderate deficits, but overall evidence is mixed and effect sizes are small. Timing, stroke severity, and session frequency influence outcomes. Limited availability, poor research evidence, and inconsistent functional benefits mean these interventions are most appropriate within research contexts rather than routine practice.

3.5.2 Existing Protocols for the Upper Limb

Standardized protocols exist to guide the rehabilitation of upper limb function after stroke. They help ensure consistency in clinical practice, promote optimal recovery outcomes, and facilitate the translation of research evidence into day-to-day rehabilitation. Several established programs and resources provide practical models for implementing these principles in diverse clinical and home-based settings (Fig. 3.1).

Motor protocols

- GRASP (Graded Repetitive Arm Supplementary Program) Protocol: The GRASP protocol is a task-specific upper limb training program designed for patients with post-stroke arm impairments in the months following stroke. The program is structured into three levels, corresponding to the severity of the patient's arm impairment (GRASP I Neurorehabilitation Research Program)
- PUSH (Promoting Use of Shoulder and Hand) programme: PUSH is an exercise program designed to help stroke survivors activate the muscles (PUSH)
- The StrokeEd MOOC (Massive Open Online Course): The course is designed for therapists and provides a structured approach to analyzing activity limitations in the upper limb, while outlining the current evidence base for post-stroke rehabilitation (The StrokeEd Collaboration conducts stroke rehabilitation workshops for physiotherapists and occupational therapists)
- Progressive resisted exercises using 1 Repetition Maximum (RM) (Prescribing strength training for stroke recovery: a systematic review and meta-analysis of randomised controlled trials I British Journal of Sports Medicine).

Sensory protocols

- SENSe protocol (Study of the Effectiveness of Neurorehabilitation on Sensation): is a neuroscience-based sensory rehabilitation program that uses principles of perceptual learning to help stroke survivors regain functional touch in the upper limb. It focuses on retraining sensory discrimination and applying these skills to everyday tasks (SENSe Therapy - SENSe)
- SMULS (Sensorimotor Exercise Regime for Upper Limb Deficits in the In-Hospital Phase After a Supratentorial Stroke): s a rehabilitation protocol designed to address both sensory and motor deficits of the upper limb following stroke, integrating sensory retraining with task-specific motor practice to enhance functional recovery. (Effect of a Sensorimotor Exercise Regime for Upper Limb Deficits in the In-Hospital Phase After a Supratentorial Stroke (SMULS)—Protocol for a Randomized Controlled Trial - Dorcas B.C. Gandhi, Nistara Singh Chawla, Christy John, Amarjot Gill Singh, Sania Zarreen, Pranay J. Vijayanand, 2023)

Fig. 3.1 Standardized protocols for upper limb rehabilitation

3.6 Models of Rehabilitation for the Upper Limb in Low-Resource Settings

In many low-resource environments, access to specialized rehabilitation services for stroke survivors remains limited (Fig. 3.2). To overcome these challenges, contextualized rehabilitation programs have been developed that adapt evidence-based practices to local realities. These programs emphasize community participation, cost-effective approaches, and culturally relevant strategies to enhance accessibility and sustainability of post-stroke upper limb recovery. Models that can be adapted include:

(a) Culturally Relevant Interventions.
Programs for rehabilitation must consider the socioeconomic and cultural realities of the neighbourhood. Patients depend on community support and family caregivers in many low-resource environments (Fig. 3.3b). Including community health workers and caregivers in treatment plans can improve results and adherence [8]. For instance, training caregivers on the fundamentals of therapeutic exercises enables them to support regular at-home rehabilitation.

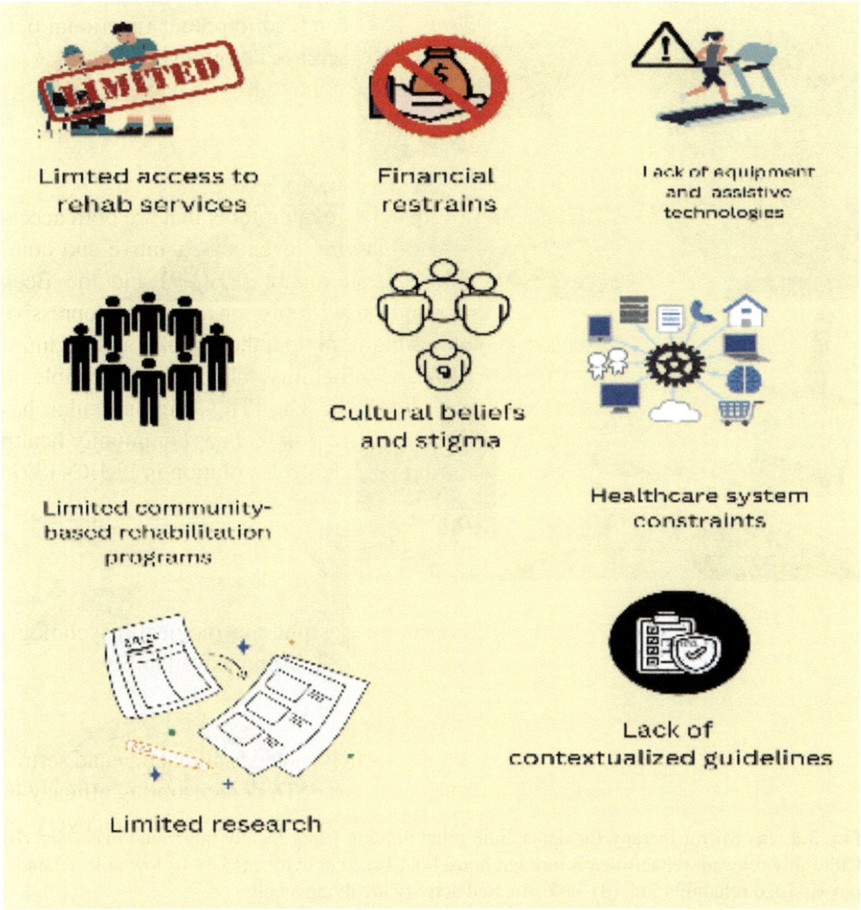

Fig. 3.2 Challenges in low-resource settings

(b) Task-Oriented Training.
Task-oriented training, which involves practicing functional activities, is effective for arm and hand recovery [9]. Therapists can use locally accessible resources in low-resource settings, such as household items for grip-strengthening exercises, in place of pricey equipment (Fig. 3.3a, d). This method lowers expenses while simultaneously making the training more applicable to patients' everyday lives [10].

(c) Use of Digital Technologies.
In low-resource environments, digital health tools such as tele-rehabilitation platforms and mobile applications are increasingly being adopted to extend access to rehabilitation services. These technologies facilitate remote delivery of therapy, monitor patient progress, and provide motivational feedback to support continued engagement. Both advanced and simplified digital solutions are

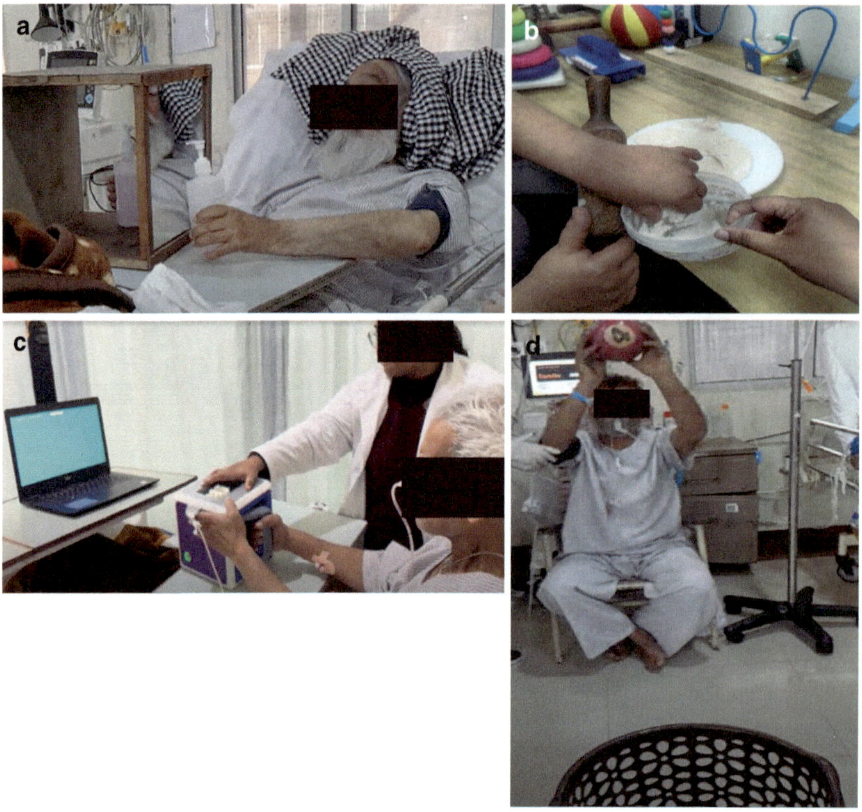

Fig. 3.3 (**a**) Mirror therapy for upper limb rehabilitation using locally fabricated materials. (**b**) Culturally relevant rehabilitation through household-based activity. (**c**) Use of low-cost, technology-assisted rehabilitation. (**d**) Task-oriented activity involving a ball

being implemented across South and Southeast Asia to enhance stroke recovery outcomes [11]. Similarly, programs such as CARE4STROKE, which integrates caregiver-mediated exercises with e-health support for early supported discharge, and the ATTEND-2 app-based tele-rehabilitation platform in India represent contextualized innovations tailored for use in LMICs.

(d) Low-Cost Robotics in LMICs.

The use of rehabilitation robotics holds promise for reducing healthcare disparities by enhancing access to consistent and intensive therapy. However, its availability in low- and middle-income countries (LMICs) remains limited due to high production and maintenance costs, thus restricting the number of patients who can benefit from such technologies (Fig. 3.3c). The development of low-cost robotic systems can address this gap through several strategies:

1. Designing end-effector and exoskeleton-based robots to target proximal and distal joints, respectively.

2. Minimizing the number of degrees of freedom (DOF) to enable effective rehabilitation within a simplified planar workspace.
3. Using lightweight, affordable materials to reduce bulk and cost, and improve portability.

For example, the PLUg and Train Robot (PLUTO) is a single DOF robot with interchangeable attachments to train multiple joints sequentially, supporting active and assisted modes with game-based feedback to motivate patients. The Arm Rehabilitation Robot (AREBO) is an end-effector robot that offers single- and multi-joint-assisted training, is compact and portable, and features optimized link lengths, with three actuated and three unactuated degrees of freedom. It delivers unassisted or adaptive weight-supported training using a human arm model without requiring additional sensors. These platforms illustrate how affordable, adaptable robotics can enhance rehabilitation in diverse settings.

(e) Community-Based Rehabilitation (CBR).

CBR programs provide rehabilitation services through local networks. These initiatives, which decentralize care to community health workers, are especially successful in addressing stroke-related dysfunction in underprivileged areas [8]. Additionally, CBR promotes inclusivity by integrating social participations, livelihood support, education, and rehabilitation [12].

3.6.1 Challenges and Recommendations

Lack of infrastructure, insufficient funding, and a lack of understanding of the significance of rehabilitation are major obstacles. Prioritizing training for local health workers, purchasing reasonably priced assistive technology, and encouraging research to assess culturally tailored interventions are all things that governments and nongovernmental organizations should do. Contextualized arm and hand rehabilitation programs can greatly enhance results in low-resource environments by utilizing community resources, incorporating cultural customs, and implementing cutting-edge technologies.

3.7 Contextualized Upper Limb Rehabilitation in LMICs

Contextualizing interventions for stroke survivors is a multifaceted challenge. Lifestyle diversity, occupational demands, socioeconomic conditions, and individual variations in physical and psychological status add layers of complexity to rehabilitation efforts. For individuals and families with limited financial resources, access to continuous rehabilitation services may not be a priority, leaving their needs unmet. In such contexts, interventions must emphasize self-training at home or in the workplace, supplemented by initial clinical training supervised by rehabilitation professionals. Exercises should be participant-centered, directly addressing daily challenges such as self-care, transportation, and work-related tasks.

To optimize effectiveness, exercise training and real-world practice must be closely integrated. Caregivers play a crucial role in this integration, but their involvement must be carefully balanced. Over-support may foster dependency, while insufficient support may hinder progress. Training caregivers to provide encouragement and graded assistance without undermining autonomy is therefore essential for sustained recovery and motivation.

The scientific rationale for bilateral arm functional training lies in its ability to normalize interhemispheric disinhibition, promote neuroplasticity in the lesioned hemisphere, and address the "nonuse phenomenon" of the more affected limb. Both gross motor and fine, task-based exercises contribute to improved adherence and mental health. Importantly, bilateral training is inherently task-specific and task-oriented, bridging structured rehabilitation with functional daily activities and offering a cost-effective, scalable strategy in LMICs.

Conceptualizing activity training for stroke survivors, certain prerequisites must be ensured to optimize outcomes. Based on the activity preferences and goals of the patients, training and practice should be initiated, incorporating fine hand motor skills such as transferring objects between the fingers, finger to palm movements, and inter-hand manipulation skills. Additionally, grasping and prehension should be practiced using objects of various sizes and textures to ensure progressive skill development.

3.7.1 Gross Movement

For gross motor movements, the "table wrap" approach can be employed to cover the full range of motion. This ensures that the skills practiced during training are directly transferable to real-world functional activities. The table wrap technique is particularly beneficial for bilateral arm functional training, as it encourages the coordinated use of both arms, promoting neuroplasticity and functional recovery (Fig. 3.4).

A table wrap, as illustrated above, is positioned in front of the patient and serves as a reference for performing bilateral activities. These activities are carried out at various angles, as depicted, with simultaneous movements of both the right and left sides. This approach aims to enhance bilateral inter- and intra-limb coordination, which in turn initiates and progressively improves movements in the frontal plane. Additionally, the straight lines marked at 1/3, 2/3, and full arm distance (measured from the acromion process to the mid-palmar crease) are used to guide activities that initiate and advance movements in the sagittal plane.

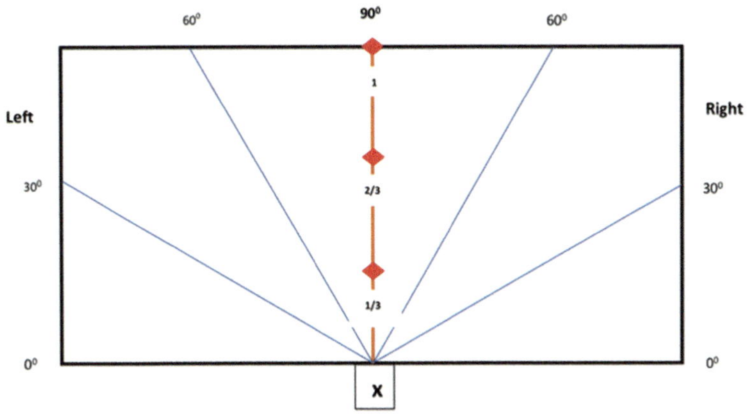

Fig. 3.4 Table wrap for bilateral arm functional training

Training can be structured in two complementary parts:

- Task-specific practice under clinical supervision, emphasizing fine motor coordination, in-hand manipulation, and targeted range of motion.
- Task-oriented practice at home, supported by caregivers and using readily available community objects to train hand functions in meaningful contexts.

During practice, progression could be attained in terms of completing the ranges of extent in all the cardinal planes and temporal parameters. Key movements such as elbow extension, forearm pronation, and wrist and finger extension could be emphasized during practice.

In Table 3.3, the mentioned task-specific activities are designed to improve fine motor skills, coordination, and strength. These activities can be seamlessly translated into task-oriented activities that align with daily living needs, ensuring functional relevance.

Training can be provided regardless of hand dominance, with the primary goal of enhancing manipulative skills in the affected hand to improve functional abilities. To ensure effective practice and progression, practice should begin with simple tasks tailored to the patient's current abilities and priorities. This approach helps build interest and motivation. Gradually, transition to training gross hand functions, followed by prehension skills. Specifically, start with symmetrical hand practice to establish a foundation, and then progress to asymmetrical practice to further challenge and refine coordination and control (Fig. 3.5).

Table 3.3 List of contextualized task training for upper limb rehabilitation in stroke survivors

Task-specific activities	Task-oriented activities
Flipping cards	Flipping newspaper pages/counting currency notes
Tying ropes	Tying knots in saree pallu, dhoti, or shoelaces
Clay-splitting activity	Breaking rotis, tearing bread pieces for dipping
Paper tearing	Tearing rotis or chapatis or banana leaves for serving food
Opening and closing a bottle cap	Opening and closing water bottles, oil containers, or pickle jars
Stacking and unstacking a plastic tray	Arranging and removing vessels in the kitchen rack
Paper folding and unfolding	Folding clothes, bedsheets, or towels
Coloring	Holding a pen and writing
Marble transfer	Transferring vegetables in a tray, transferring grains or pulses from one container to another
Paper punching	Using a stapler or hole puncher for documents
Locking and unlocking	Locking and unlocking almirahs, doors, or tiffin boxes
Bean filling	Filling spice jars, pulses, or rice containers
Clay cutting	Cutting vegetables
Stacking and unstacking the glasses	Arranging steel tumblers or cups in the kitchen
Emptying beans from a bottle	Pouring rice or dal from a container into a cooking pot
Scooping and filling	Serving rice, dal, or curry with a spoon
Clamping	Using tongs to flip dosas or pick up hot chappatis
Dropping a ball	Dropping vegetables into a cooking pan
Erasing	Cleaning a kitchen counter, wiping a slate, white board
Line drawing	Drawing rangoli or kolam patterns
Paper cutting	Cutting vegetable packets, or newspapers
Shoebox lifting	Lifting grocery bags, small containers, or vegetable baskets
Folding towel	Folding sarees, dhotis, handkerchiefs
Zipping and unzipping	Zipping a purse, handbag, or luggage bag
Buttoning and unbuttoning	Buttoning shirts, kurtas, or school uniforms
Net and bolt activity	Tightening or loosening gas stove knobs or pressure cooker lids

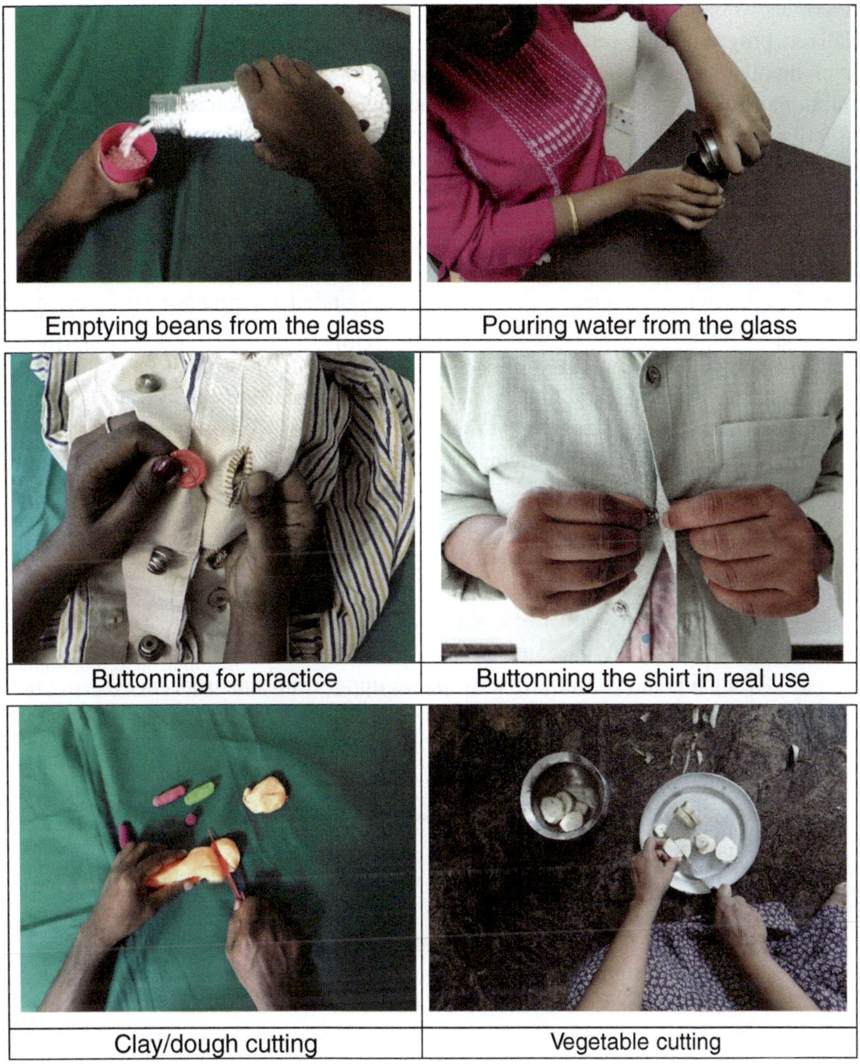

Fig. 3.5 Contextualized upper limb activities during task practice and real-life application

3.8 Summary

Stroke-related upper limb impairment is a major cause of long-term disability, particularly in LMICs, where rehabilitation resources are limited. Recovery is often slow, incomplete, and complicated by secondary issues such as shoulder subluxation and CRPS. Evidence-based therapies, including CIMT, mirror therapy, mental practice, progressive resistance training, and task-specific training, can improve outcomes when adapted to patient needs. In resource-constrained settings, contextualized programs that integrate culturally relevant, task-oriented activities, caregiver involvement, and community-based rehabilitation are crucial. Bilateral arm functional training, supported by simple household-based exercises, offers a cost-effective, scalable strategy to enhance independence and participation.

What Does Evidence Say

- Progressive resistance training is recommended to improve strength in stroke survivors with limb weakness.
- Repetitive, task-specific training should be used to improve arm and hand function when some voluntary movement is present.
- Constraint-induced movement therapy (CIMT), including intensive protocols, is effective for improving arm and hand use in patients with active wrist and finger extension.
- Assistive technologies—including robotics, virtual reality, and interactive gaming—may be used to enhance upper limb function.
- Electrical stimulation may be used:
  - To improve strength in patients with significant weakness (especially below antigravity strength).
  - Alongside motor training to improve upper limb function.
  - Early after stroke (within 6 months) to prevent or reduce shoulder subluxation.
  - To help manage shoulder pain.
- Sensory-specific training is recommended for individuals with upper limb sensory loss.
- Mental practice and mirror therapy may be used as adjuncts to improve arm function, particularly in patients with mild-to-moderate weakness.
- Mirror therapy may also help improve function and daily activities in patients with unilateral neglect.
- Botulinum Toxin A, combined with rehabilitation, may reduce spasticity, though it is unlikely to significantly improve functional outcomes.
- Adjuncts such as electrical stimulation, casting, or taping may be used alongside Botulinum Toxin A.
- Routine stretching, splinting, or acupuncture is not recommended for reducing spasticity unless in a research setting.
- Education on proper positioning and handling should be provided to patients, caregivers, and healthcare staff to prevent complications.
- Electrical stimulation may help prevent or reduce shoulder subluxation early after stroke.
- Shoulder strapping is **not recommended** for preventing subluxation.

(continued)

- Routine use of splints or stretching to prevent contractures is **not recommended** in patients receiving active rehabilitation.
- Shoulder strapping, electrical stimulation, and appropriate injections (e.g., steroid or nerve block) may be used to reduce pain.
- Botulinum Toxin A may help reduce pain when spasticity is present.
- Acupuncture may be considered as an adjunct to rehabilitation for pain relief.
- Hand and wrist orthoses (splints) do not improve function, pain, or range of motion and **should not be used routinely**.
- Brain stimulation techniques (tDCS, rTMS) **should not be used outside research settings**.
- Acupuncture for spasticity and routine stretching for spasticity prevention are **not recommended** in standard practice.

References

1. Katan M, Luft A. Global burden of stroke. Semin Neurol. 2018;38:208–11. https://doi.org/10.1055/s-0038-1649503.
2. Kwakkel G, Stinear C, Essers B, Munoz-Novoa M, Branscheidt M, Cabanas-Valdés R, Lakičević S, Lampropoulou S, Luft AR, Marque P, Moore SA, Solomon JM, Swinnen E, Turolla Andrea, Murphy MA, Verheyden G. Motor rehabilitation after stroke: European Stroke Organisation (ESO) consensus-based definition and guiding framework, European Stroke Journal. 2023;8(4):880–94. https://doi.org/10.1177/23969873231191304.
3. Doumen S, Sorba L, Feys P, Tedesco Triccas L. Efficacy and dose of rehabilitation approaches for severe upper limb impairments and disability during early acute and subacute stroke: a systematic review. Phys Ther. 2023;103:pzad002. https://doi.org/10.1093/ptj/pzad002.
4. Bernhardt J, Hayward KS, Kwakkel G, Ward NS, Wolf SL, Borschmann K, Krakauer JW, Boyd LA, Carmichael ST, Corbett D, Cramer SC. Agreed definitions and a shared vision for new standards in stroke recovery research: the stroke recovery and rehabilitation roundtable taskforce. Int J Stroke. 2017;12:444–50. https://doi.org/10.1177/1747493017711816.
5. Levin MF, Kleim JA, Wolf SL. What do motor "recovery" and "compensation" mean in patients following stroke? Neurorehabil Neural Repair. 2009;23:313–9. https://doi.org/10.1177/1545968308328727.
6. Raghavan P. Upper limb motor impairment post stroke. Phys Med Rehabil Clin N Am. 2015;26:599–610. https://doi.org/10.1016/j.pmr.2015.06.008.
7. Carlsson H, Gard G, Brogårdh C. Upper-limb sensory impairments after stroke: self-reported experiences of daily life and rehabilitation. J Rehabil Med. 2018;50:45–51. https://doi.org/10.2340/16501977-2282.
8. Maanoosi M. Stroke Ser South Sudan Med J. 2024;17(1):27–31.
9. Hussain M, et al. Effects of task oriented rehabilitation of upper extremity after stroke: a systematic review. Jof Pak Med Assoc. 2022;72(7):1406–15. https://doi.org/10.47391/JPMA.3864.
10. Winstein CJ, et al. Guidelines for adult stroke rehabilitation and recovery: a guideline for healthcare professionals from the American Heart Association/American Stroke Association. Stroke. 2016; https://doi.org/10.1161/STR.0000000000000098.

11. Selamat SNS, et al. The application of technological intervention for stroke rehabilitation in Southeast Asia: a scoping review with stakeholders' consultation. Front Public Health. 2022;9:1–12. https://doi.org/10.3389/fpubh.2021.783565.
12. World Health Organization. Community-based rehabilitation: CBR guidelines. Geneva: World Health Organization; 2010. p. 67. https://www.who.int/publications-detail/community-based-rehabilitation-cbr-guidelines

Gait and Balance Rehabilitation Post-Stroke in Low-Resource Settings

4

Gerard Urimubenshi, Anne Kumurenzi, Farayi Kaseke, Roberta de Oliveira Cacho, Jussara Almeida de Oliveira Baggio, and Dorcas B. C. Gandhi

Abstract

Stroke-related impairments of gait and balance are major contributors to long-term disability, limiting mobility, independence, and quality of life. In low-resource settings, these challenges are compounded by inadequate infrastructure, scarcity of rehabilitation professionals, socioeconomic barriers, and limited access to assistive technologies. This chapter provides a comprehensive overview of gait and balance rehabilitation post-stroke, drawing on current evidence while emphasizing context-specific solutions for resource-constrained environments.

It begins with the neurophysiological basis of gait and balance impairments and their functional and psychosocial consequences. Evidence-based rehabilitation strategies, including task-specific and repetitive practice, circuit class ther-

G. Urimubenshi (✉) · A. Kumurenzi · F. Kaseke
Department of Physiotherapy, College of Medicine and Health Sciences, University of Rwanda, Kigali, Rwanda
e-mail: g.urimubenshi@ur.ac.rw; a.kumurenzi@ur.ac.rw; f.kaseke@ur.ac.rw

R. de Oliveira Cacho
Faculty of Health Sciences at Trairi (FACISA), Federal University of Rio Grande do Norte (UFRN), Santa Cruz, Rio Grande do Norte, Brazil

J. A. de Oliveira Baggio
Medical Sciences and Nursing Complex – CCME, Federal University of Alagoas, Arapiraca, Alagoas, Brazil

D. B. C. Gandhi
Department of Neurology, Christian Medical College & Hospital Ludhiana, Ludhiana, Punjab, India

Department of Physiotherapy, Manipal Academy of Higher Education, Manipal, Karnataka, India

D. B. C. Gandhi, J. D Pandian (eds.), *Stroke Rehabilitation in Low-Resource Settings*, https://doi.org/10.1007/978-981-92-0403-8_4

apy, treadmill training, virtual reality, functional electrical stimulation, and strength training, are reviewed, with an emphasis on their adaptability using low-cost materials and community facilities. Practical approaches for assessment, goal setting, dosage, safety, and integration of evidence into daily rehabilitation are discussed, highlighting the creative use of household resources, caregiver support, and group-based interventions.

The chapter also examines barriers to effective rehabilitation, including limited service availability, cultural perceptions of stroke, financial constraints, and environmental hazards. To address these gaps, it outlines capacity-building priorities such as tele-rehabilitation, task-shifting to community health workers, training programmes, and the development of national rehabilitation guidelines. Policy and health system implications are underscored, advocating for rehabilitation to be recognized as a public health priority with sustained funding and universal health coverage.

Future directions call for scalable, low-cost, and culturally relevant interventions, strengthened community-based rehabilitation, and innovative digital technologies to improve accessibility and adherence. By combining evidence-based practice, local adaptation, and system-level support, gait and balance rehabilitation can significantly enhance functional recovery and reintegration of stroke survivors in low-resource contexts.

Keywords

Stroke • Gait • Balance • Rehabilitation • Low-resource settings • Community-based rehabilitation • Tele-rehabilitation

4.1 Introduction

Stroke significantly impairs gait and balance, which are critical for functional recovery and quality of life, making it a leading cause of disability globally [1]. After a stroke, gait and balance deficiencies often result in reduced mobility, an increased risk of falls, and a permanent dependence on caregivers [2]. Effective rehabilitation is necessary to restore these functions, but it is still difficult to provide evidence-based gait and balance rehabilitation in low-resource settings where access to healthcare, trained staff, and rehabilitation technologies is limited [3].

Stroke survivors' recovery is hampered in many low- and middle-income countries (LMICs) by systemic obstacles, like poor infrastructure, limited funding, and a lack of knowledge about rehabilitation services [4]. Furthermore, these issues are made worse by the lack of physiotherapists and specialized equipment, which leaves significant portions of the population underserved [5]. Innovative, scalable, and economical strategies that are adapted to the circumstances of resource-constrained environments are needed to overcome these obstacles.

Important facets of post-stroke gait and balance rehabilitation in low-resource environments are examined in this chapter. After giving a summary of the gait and balance issues that stroke survivors frequently experience, it analyses the obstacles that prevent them from accessing rehabilitation services. Task-oriented exercises, community-based rehabilitation (CBR), and low-cost technology solutions are examples of evidence-based interventions that are discussed as workable ways to enhance results in environments with limited resources.

The chapter also discusses capacity-building tactics to improve rehabilitation services, like tele-rehabilitation, training regional healthcare providers, and creating supportive policy frameworks. To guarantee fair and long-lasting rehabilitation access, future directions and research priorities are finally described, highlighting the necessity of context-specific solutions, cutting-edge technologies, and stakeholder collaboration.

To enhance recovery as well as life after stroke, this chapter attempts to give a thorough grasp of the opportunities to enhance gait and balance rehabilitation after a stroke in low-resource settings.

4.2 Gait and Balance Impairments Post-Stroke

Because the brain's motor control pathways are disrupted after a stroke, the neurophysiology of gait and balance is greatly impacted. The brain is essential for controlling the movement patterns needed for walking and balance, especially the motor cortex, basal ganglia, and cerebellum [6]. These brain regions are frequently damaged during a stroke, resulting in motor impairments like spasticity or hemiparesis (partial paralysis), which significantly impair a person's capacity to walk and maintain balance [7].

To regulate lower limb movements and maintain postural stability and coordinated actions, the central nervous system (CNS) combines sensory input from the vestibular, somatosensory, and visual systems [6]. A stroke, on the other hand, impairs brain-muscle communication, resulting in aberrant gait patterns like asymmetry, shortened strides, and decreased speed [7].

Stroke patients frequently have compromised corticospinal tracts, which carry motor signals from the brain to the muscles. This disturbance makes it harder to voluntarily control the leg muscles, which makes it harder to start and continue walking. The basal ganglia, which control movement, may also be impacted, which could lead to problems like bradykinesia (slowness of movement) or rigidity [8].

Balance is also disrupted by the stroke's impact on the brain's ability to integrate and process sensory information [1]. Postural instability may result from vestibular dysfunction and a loss of proprioception or the awareness of one's own body position. It may also be more difficult for stroke survivors to recover from a loss of balance or correct postural sway due to impairments in the cerebellum, which coordinates fine motor control and balance [6]. To improve gait and balance in stroke survivors, rehabilitation strategies such as physical therapy and neuromodulation techniques work to increase neuroplasticity, which aids in the brain's reorganization and compensation for lost function [7].

One of the most prevalent and incapacitating effects of stroke is impaired gait and balance, which has a major negative influence on mobility and general quality of life. Damage to motor control, sensory processing, and coordination following a stroke frequently results in asymmetrical gait patterns, decreased walking speed, and impaired balance, all of which reduce functional independence. Mobility is especially impacted because people with gait and balance disorders have difficulty carrying out daily tasks like climbing stairs, walking indoors, and negotiating uneven ground [1]. Because of their limited mobility, they are less able to engage in social and community activities, which increases their reliance on caregivers and assistive technology. Many stroke survivors experience early fatigue because of their decreased walking efficiency, which makes it harder for them to perform rehabilitation exercises or other physical activities that are essential to their recovery.

The consequences significantly affect life quality and go beyond physical restrictions. Loss of autonomy, low self-esteem, and increased emotional distress, such as anxiety and depression, are frequently linked to reduced mobility [9]. People with balance issues often worry about falling, which makes it harder to engage in social and physical activities. This leads to a vicious cycle of social isolation and physical inactivity [2]. Additionally, poor gait and balance can raise the chance of falls, which can lead to further functional decline, hospitalizations, and injuries. In addition to burdening stroke survivors, these negative consequences put a heavy financial and psychological burden on caregivers and healthcare systems.

Enhancing mobility, regaining functional independence, and improving quality of life all depend on treating gait and balance impairments with rehabilitation interventions like physiotherapy, gait training, and cutting-edge technologies [9, 10]. Stroke survivors can increase their participation in meaningful life activities, regain their confidence, and lower their risk of falling by improving these impairments.

4.3 Approaches for Gait and Balance Rehabilitation Post Stroke

4.3.1 Key Overarching Principles

1. Recovery is driven by neuroplasticity and repetition, task specificity, and progression/challenges are crucial.
2. Early initiation is essential, as soon as medical stability allows.
3. Progression (i.e., increasing difficulty, speed, complexity) is invaluable.
4. Balance training needs to be initiated as early as possible during the in-hospital phase; without it fall risk increases.
5. Combine multiple approaches like strength training, balance training, motor control, and assistive devices where needed.
6. Tailor therapy to the patient's mobility status (non-ambulatory, assisted ambulation, independent but impaired) and recommended assessments.

4.3.2 Recommended Assessments

At a minimum a comprehensive assessment should include evaluation of (a) impairments/body structure and function levels (muscle strength, especially lower limb/hip/ankle, spasticity/tone, sensation/proprioception, range of motion, joint restrictions, trunk control/core stability, motor control, selective movement, reaction/postural responses), (b) activity limitation (static and dynamic balance, gait (all gait parameters: spatial and temporal, speed, endurance, symmetry, variability, transfers (sit-to-stand, stand-to-sit), dual-task walking (if relevant), obstacles, stairs, turning), (c) participation restriction (falls history or risk, walking for community (distance, terrain), use of assistive devices, patient's perceived mobility using PROMs (patient-reported outcome measures)), (d) prognostic metrics (early gait speed, early balance metrics, self-efficacy, cognitive status, comorbidities), and (e) safety metrics (fall risk scales, cardiovascular tolerance).

The International Stroke Recovery and Rehabilitation Alliance, World Stroke Organization, and current evidence have recommended a few consensus based-assessment measures for the post-stroke phases, as enumerated below:

(a) Body function and structure: Fugl-Meyer Assessment—lower limb (suitable from acute to chronic phases post-stroke).
(b) Activity limitation: 10-meter walk test (timed or untimed) or 6-minute walk test limb (suitable from acute to chronic phases post-stroke).
(c) Global disability: modified Rankin score (suitable for early subacute to chronic phases post stroke).
(d) Fall risk: Falls Efficacy Scale, Berg Balance scale, Postural Assessment Scale for Stroke.
(e) Trunk Control: Trunk Control test, Trunk Impairment scale.
(f) Others: Timed Up and Go test, Instrumented measures with wearable sensors (not mandatory).

4.3.3 Specific Interventions

From recent evidence and guidelines [11–13], we have summarized the interventions that can be easily implemented in low-resource settings with minimal infrastructure and workforce. It is crucial that all stroke survivors who experience difficulty walking be supported to engage in tailored, repetitive practice of walking or its components as much as possible (Table 4.1).

Circuit class training (CCT) (Fig. 4.1) is an evidence-based, resource-efficient model mainly focused on increasing scheduled therapy time, and overground walking practice is crucial for gait recovery. In CCT, small groups of patients rotate through task-specific stations, such as overground walking, sit-to-stand, step-ups, balance exercises, and walking to a metronome (Fig. 4.2). It has been shown to be both safe and effective in increasing walking distance and speed while maximizing therapist efficiency [14]. Current evidence shows that CCT significantly improves

Table 4.1 Decision making tips

Phase	Focus	Assessment battery	Interventions/strategy
Early/acute (once medical stability)	Prevent secondary complications, early mobilization, begin neuromotor re-education	Basic neurological exam, trunk control, PASS or sitting/standing balance, muscle tone/strength, 10 MWT (if safe), TUG (if safe)	Seated balance training, weight shifts, trunk control, stepping-in-place, pre-gait activities, active/assisted movement, neuromodulation (if available)
Subacute (weeks to months)	Reinforce gait, balance, endurance, confidence	Full battery: 10 MWT, 6 MWT, BBS/FGA, TUG, strength tests, balance perturbation tests	Task-specific gait, treadmill + overground, strength training, balance exercise (perturbations, dynamic), FES, biofeedback, VR adjuncts
Transition to community/outpatient/chronic	Increase complex walking (terrain, dual-task), endurance, independence	Re-measure battery plus dual-task walking, obstacle courses, community walk test	Community-based walking, stair negotiation, ambient challenges, progression of intensity, fall prevention training
Maintenance/long term	Prevent decline, maintain gains, adapt to aging/comorbidities	Periodic reassessment (6–12 months) of core measures	Home exercise programs, periodic supervised sessions, feedback/tele-rehab, adaptive devices if needed

walking distance (6 MWT) and gait speed (10 MWT) compared to standard physiotherapy, although improvements in balance outcomes are less consistent. Trials report CCT as safe, with very few adverse events, and particularly effective for stroke survivors who already have some ambulatory capacity. The programme's effectiveness depends on progression, with tasks becoming more challenging through changes in speed, obstacle negotiation, or dual-tasking to drive neuroplasticity. CCT maximizes active therapy time and has been shown to reduce inpatient length of stay in some studies. However, most evidence comes from subacute and chronic stroke populations, with less data available for very acute or non-ambulatory patients. Long-term retention of gains and standardized protocols also remain areas that need further study. Overall, CCT is an evidence-based, efficient, and scalable approach to improving mobility after stroke when delivered with progression and appropriate safety measures. In South Africa, caregiver-assisted, task-oriented CCT focusing on overground walking significantly improved gait and balance [15]. Similarly, treadmill training, with or without body weight support, has demonstrated improvements in walking speed and endurance [16], particularly among individuals who are already able to walk at the start of treatment.

Practical tips for applying CCT in stroke gait rehab

- Start with a small-scale setup: Try with a small group (4–5 stroke survivors), 2–3 trial stations, to optimize transitions.
- Match participants by functional level or have stations ranging from easy to hard tasks.

Fig. 4.1 Circuit class training for post-stroke gait rehabilitation

- Prioritize walking-related tasks like overground walking, stepping over obstacles, and changing directions to a cue. Do not limit activities to strength or conditioning training alone.
- Include variability like dual tasks (carry object, talk while walking), tasks on uneven terrain, quick turns, pacing with a metronome, step-ups, and reactive stepping.
- Track active time by recording the time during which the patient actually engages in the task, with an aim to maximize it.
- Safety is a priority and can be ensured using gait belts, spotters, parallel bars, or supports as needed. Make sure to screen for risk factors on entry to the programme.
- As patients improve, they should progress to more challenging tasks and reduce assistance.
- Embed home-oriented "carry-over" tasks like community ambulation, stair climbing, etc.

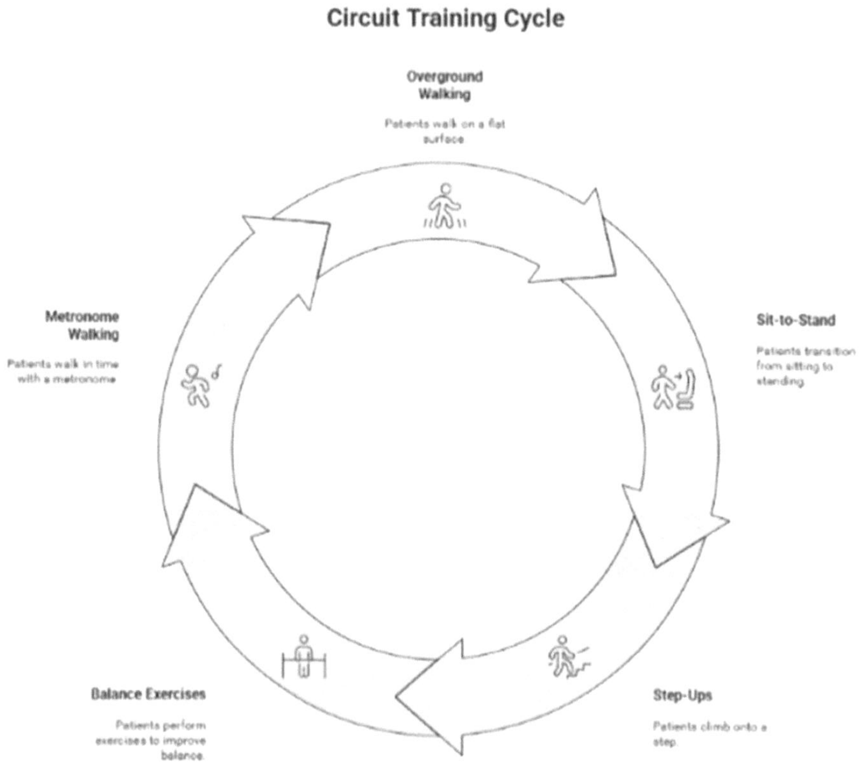

Fig. 4.2 An example of CCT

- Measure outcomes reliably and periodically with 10 MWT/6 MWT/TUG/balance scales with adequate follow ups after the programme ends.
- Monitor adverse effects like fatigue, soreness, and have criteria to scale back when/if needed.
- Consider hybrid or mixed delivery models where there is a mix of sessions in circuit mode, and 1:1 or pair work for complex/difficult tasks.

Strength Training Beyond gait-specific approaches, strength training is a key intervention for stroke survivors with reduced muscle strength. A recent meta-analysis highlighted the potential of resistance training to improve comfortable walking speed, although its impact on fast walking and endurance remains inconclusive [17]. Strengthening programmes, including core stability training, have shown benefits for balance and gait performance [18, 19]. For patients who struggle to move from sitting to standing, task-specific training such as structured sit-to-stand practice, stair climbing, and balance exercises can improve functional independence [19], while repeated practice of standing up has been shown to be effective in community-based rehabilitation in Nigeria [20]. Evidence highlights the need for progressive resistance training, especially for the hip extensors, knee extensors, and ankle plantarflexors/dorsiflexors. Lower limb weakness is one of the strongest predictors of impaired walking after stroke [21]. Progressive resistance training (PRT) improves strength and translates into functional walking gains, with no evidence indicating that it worsens spasticity [22]. Training protocols are recommended to:

(a) Follow progressive overload and task relevance.
(b) Use multi-joint, functional movements (sit-to-stand, step-ups, resisted walking) in addition to isolated resistance exercises.
(c) Prioritize lower-limb strength with inclusion of core and trunk musculature for postural stability.

Key evidence for strength training post-stroke states the need for integrating gait training along with strength training for improved speed, distance coverage, and functional ambulation. AHA/ASA guidelines, in fact, recommend combined strength, aerobic, and task-specific mobility training for best gait outcomes. The dosage needs to be tailor-made to every stroke survivor, but 60–80% 1RM, 2–3 times per week with appropriate warm-up and cool down periods is recommended. Cool down periods could include short overground walking or treadmill intervals to reinforce motor carryover [11, 23, 24].

Aerobic Training Current evidence also recommends aerobic training right from the acute phase post-stroke (see guidelines provided in the panel at the end of the chapter). Such training increases cardiovascular reserve and fatigue tolerance, which subsequently enables higher doses of task-specific practice—a key driver for gait recovery. Clinically, the mantra of screen-prescribe-progress works well for aerobic training. Most studies recommend 3–5 sessions/week for over 8 weeks to see any significant changes in outcomes, with typically 40–80% heart-rate reserve (HRR) or 50–80% HRmax, or RPE 11–16 [25] with each session lasting for 20–40 min (continuous or with intervals based on patient status) [26]. Integrating task-specific gait training with aerobic training within the same programme is essential for better functional transfer [26]. Such a programme should be implemented carefully with screening for cardiac risk, uncontrolled hypertension, orthopaedic limits, autonomic instability, and severe cognitive/language issues that limit safe participation (ECG and cardiology clearance essential). Additionally, always start with supervised sessions (BP, HR, RPE). A quick checklist for aerobic training could be as follows:

(a) Screen and clear for any medical status influencing aerobic training.
(b) Screening for cognition to ensure the ability to follow commands and report adverse effects, if any.
(c) Initiate with three 20-min aerobic (moderate intensity: 40–60% HRR) per week, progressing to 4–5 sessions per week, and 30–40 min in duration, with higher intensity as tolerated.
(d) Incorporate walking tasks. If this is unsafe, use cycling/recumbent stepping.

(e) Combine aerobic training with gait drills and strength training.
(f) Record VO_2 surrogate (6 MWT) and gait speed at baseline and periodically to track benefit.

Alongside these core approaches, additional modalities can be integrated to further improve walking outcomes. Virtual reality (VR) training, for instance, has been trialled in several LMICs, including VR-assisted gait adaptation training in China [27] and VR combined with Lokomat therapy in Turkey [28], both of which showed promise for enhancing walking recovery. Electromechanically assisted gait training, such as Lokomat-based interventions in Turkey [29], and biofeedback-supported training, including robotic gait orthoses with synchronized treadmill training [30], have also reported positive outcomes. Evidence suggests that electromechanical-assisted gait training combined with physiotherapy increases the likelihood of achieving independent walking compared to conventional gait training, particularly for those in the first 3 months post-stroke or for individuals unable to walk at baseline [31]. Other complementary approaches include walking training with cueing of cadence, which has been shown to improve walking speed and stride length when added to conventional therapy [32], and electrical stimulation, particularly functional electrical stimulation (FES). FES has proven effective in improving mobility, balance, gait performance, and reducing spasticity [33], with ankle dorsiflexion range and motor function improving significantly in patients with foot drop during the recovery phase [34].

For survivors with impaired standing ability or balance deficits, combining practice with adjunct technologies can further enhance outcomes. VR-based interventions [35], exergame-based training [36], and motion-capture or feedback devices such as smartphone-linked balance discs [37] have all been trialled with success. Electromechanically assisted standing or gait training [38] and visual or auditory feedback through devices such as robotic orthoses or force platforms [30] also show promise.

Task-Specific Task-specific training remains fundamental for improving balance, and studies have demonstrated the benefits of individualized core stability training [39] as well as proprioceptive and balance strategy training [40, 41]. In addition, multifactorial community-based interventions targeting fall risk have been implemented successfully, for example, through self-management and education programmes in Nigeria [42]. Supplementation should also be considered for stroke survivors with severe vitamin D deficiency, given evidence that it enhances motor function and mobility post-stroke [38]. Evidence suggests that task-specific gait practice, which is not limited to repetitive stepping, overground walking, and community-relevant tasks, is effective with task specificity and dose/intensity being the key drivers of gait recovery [43, 44]. Simple tasks, but at high intensity, can be easily implemented in LMICs with limited infrastructure/workforce. For example, high-intensity stepping practice with many steps per session, which is also taxing for the cardiovascular system, improves walking speed and endurance more than low-intensity or non-stepping training [45]. Dual-task (cognitive-motor) training and combined interventions may help real-world walking in some subgroups [46]. One important practical tip is that dose matters, so document it as total duration, number of steps, repetitions, and intensity at which the task was performed (heart rate or RPE). Higher cumulative practice (minutes and steps) correlates with better outcomes across trials and reviews.

Assistive Devices Orthotic devices can play a role in reducing mobility limitations. In Iran, ankle-foot orthoses (AFOs) designed to resist plantar flexion improved gait efficiency in stroke survivors [47], although their benefits are limited to periods when they are worn. For those facing challenges with outdoor mobility, individualized goal-setting combined with adaptive equipment, appropriate information, and referrals can support community reintegration. Escorted walking practice in real-world environments has also been recommended, and when supplemented with virtual reality that simulates community walking, as in recent trials in China [27], outcomes may be further enhanced. Finally, stroke survivors with limb apraxia may benefit from specialized approaches such as gesture training, strategy training, and errorless learning, all of which have shown promise in supporting functional recovery [48].

4.4 Practical Integration of Evidence into Daily Gait Rehabilitation

Gait rehabilitation after stroke is most effective when grounded in task-specific, repetitive, and progressively challenging practice. Systematic reviews and clinical guidelines highlight that walking and related activities should be practiced frequently, with gradual increases in speed, distance, and complexity to stimulate neuroplasticity and enhance functional recovery [11, 49, 50]. Therapy should emphasize active patient participation, with minimal passive handling [51], while safety remains a priority—particularly in settings where protective equipment is limited.

Goal setting is equally central. Evidence supports the use of SMART (specific, measurable, achievable, relevant, time-bound) goals in collaboration with patients, which improves motivation, engagement, and adherence [52, 53]. These goals should be functional and context-specific, such as enabling outdoor walking with a stick over a set distance within a defined timeframe.

In low-resource settings, as illustrated in Fig. 4.3, adaptability and creativity are essential. Simple household objects such as chairs, blocks, broom handles, water bottles, or sandbags can substitute for formal equipment, while smartphones can be used as metronomes, pedometers, or video-feedback tools. Community halls, schoolyards, and church compounds provide safe and accessible environments for group or circuit-based training [54].

Safety and real-world practice must always frame rehabilitation. Patients at risk of falls should train near stable supports such as chairs or parallel bars, and high-risk individuals should be supervised [55]. Education for patients and caregivers about reducing household hazards and installing handrails enhances home safety. To

Fig. 4.3 Example of creative use of household items and community spaces for rehabilitation in low-resource settings

ensure skills generalize, practice should extend into real-world community contexts such as markets, slopes, and stairs [56].

Finally, the application of guidelines in daily practice requires pragmatic adaptation. Walking can be structured into timed blocks with step counts recorded for feedback. Circuit class therapy principles can be applied to individual patients by rotating through tasks like sit-to-stands, step-ups, and overground walking [14, 15]. Where treadmills are unavailable, corridor walking with cadence cues provides an effective alternative [57]. Advanced technologies such as robotics or virtual reality can be replaced with low-cost strategies, including caregiver-assisted counting or metronome apps. Strength training can be integrated seamlessly into functional daily activities, such as repeated sit-to-stands, stair climbing, or carrying weighted objects [19, 20].

In sum, evidence-based gait rehabilitation can be translated into practical, safe, and adaptable strategies that are feasible across diverse resource contexts. By combining structured assessment, patient-centred goal setting, adequate dosage, creative use of available resources, and real-world practice, clinicians can maximize functional recovery and community reintegration for stroke survivors [11, 55].

4.4.1 Established and Simplified Protocols for Post-Stroke Gait/Mobility Rehabilitation

Though there is no universally accepted gold standard for post-stroke gait protocols, there are a few "package programmes", frameworks, or staged protocols that are commonly used and have a large evidence base that are suitable for implementation in LMICs. One good example is the Fitness and Mobility Exercise (FAME) protocol, which is a group-based programme for those who can stand/walk to some extent. This typical 1-h programme, suitable for community settings, includes warm-up, functional strengthening, agility/fitness, and balance and cool-down components. With a progressive intensity component, FAME targets multiple domains rather than just walking drills and has proven to improve walking speed, strength, balance, and reduce the risk of falls (https://fameexercise.com/).

Another programme is the High Intensity Gait Training (HIGH) protocol that emphasizes high step counts, high speed, perturbations, practice on varied terrains, multidirectional stimulus, and directional changes. The programme uses weights, increments in speed, and dual tasking to provide intensive, repetitive, and task-based training.

4.5 Future Directions and Research Priorities

Innovative, economical, and scalable strategies are needed for gait and balance rehabilitation in low-resource environments to address the post-stroke needs. Development and validation of cost-effective interventions for people in underprivileged communities, where access to sophisticated equipment and skilled staff may be limited, should be the top priority of future research. Using digital tools like wearable technology, tele-rehabilitation, and mobile health (mHealth) apps to deliver guided exercises, track progress, and give remote feedback is one important avenue. Particularly in remote areas with limited accessibility to healthcare, these technologies can bridge the gap between specialists and patients. Future research should evaluate these digital tools' viability, efficacy, and scalability in low-income settings.

Additionally, CBR's role needs to be strengthened and expanded. CBR provides rehabilitation services at the household or community level by utilizing caregivers, community health workers, and local resources [58].

Task-oriented, culturally relevant exercises that are simple to execute and don't require costly equipment or highly skilled professionals should be the focus of future research. It's also important to look into less expensive alternatives to sophisticated rehabilitation tools, like inexpensive robotics, adapted exercise equipment, and creative use of regional resources. Low-tech interventions that incorporate physiotherapy techniques can offer long-term alternatives for training gait and balance.

Furthermore, there is a need for more robust studies to assess the long-term impact of these interventions, especially when it comes to removing accessibility and adherence obstacles. Strategies for inclusive rehabilitation that take socioeconomic status, comorbidities, and gender into account should be prioritized. To develop evidence-based interventions and put into practice sustainable models that promote recovery as well as quality of life post-stroke in low-resource settings, cooperation between researchers, legislators, and healthcare providers will ultimately be essential.

4.6 Conclusion

After a stroke, problems with gait and balance affect mobility, functional independence, and quality of life, leading to social, emotional, and physical difficulties. Effective rehabilitation is further hampered in low-resource settings by several obstacles, such as restricted access to healthcare, a lack of qualified professionals, socioeconomic limitations, and environmental difficulties. Improving results requires removing these obstacles through creative, economical interventions, capacity building, and legislative modifications. Restoring mobility and improving the quality of life for stroke survivors in underprivileged areas may be possible through the integration of digital technologies, community-based rehabilitation, and affordable solutions.

What Does Evidence Say

- Stroke survivors with walking difficulties should perform intensive, repetitive, task-specific walking practice (or its components), using methods such as circuit class therapy (focused on overground walking) and treadmill training with or without body weight support.
- Additional interventions to support walking may include virtual reality training, electromechanically assisted gait training, biofeedback, cadence cueing, and electrical stimulation.
- Individuals with reduced strength in the arms and/or legs should undergo structured strength training to improve post-stroke muscle function.
- Those with difficulty moving from sitting to standing or rising from a chair should practice repeated, task-specific sit-to-stand activities.
- Stroke survivors with impaired standing or balance should engage in task-specific balance training with progressively challenging activities.
- Adjunct therapies for standing and balance may include virtual reality systems (e.g., treadmill-based or motion-sensing platforms), visual or auditory biofeedback (e.g., force platforms), and electromechanically assisted standing or gait training.

(continued)

- Individuals at risk of falls should receive multifactorial, community-based interventions, including personalized exercise programmes, safety advice, and vitamin D supplementation if deficient.
- Individually fitted lower limb orthoses may be used to improve walking ability, with benefits typically present only while the orthosis is worn.
- Stroke survivors with difficulty in community or outdoor mobility should set individualized goals and receive support through adaptive equipment, information, and referrals; supervised walking practice in varied environments, including virtual simulations, may help.
- For those with limb apraxia, interventions such as gesture training, strategy training, and errorless learning techniques may be provided.

References

1. Gittins M, Lugo-Palacios D, Vail A, Bowen A, Paley L, Bray B, et al. Stroke impairment categories: a new way to classify the effects of stroke based on stroke-related impairments. Clin Rehabil. 2021;35(3):446–58.
2. Gobezie M, Kassa T, Suliman J, Eriku GA, Takele MD. Balance impairment and associated factors among stroke survivors in public hospitals of Amhara regional state: a multicenter cross-sectional study. BMC Neurol. 2024;24:387–97.
3. Van Zyl C, Badenhorst M, Hanekom S, Heine M. Unravelling "low-resource settings": a systematic scoping review with qualitative content analysis. BMJ Glob Health. 2021;6(6):1–14.
4. Bernhardt J, Urimubenshi G, Gandhi DBC, Eng JJ. Stroke rehabilitation in low-income and middle-income countries: a call to action. Lancet. 2020;396(10260):1452–62.
5. Pandian JD, Kalkonde Y, Sebastian IA, Felix C, Urimubenshi G, Bosch J. Stroke systems of care in low-income and middle income countries: challenges and opportunities. Lancet. 2020;396(10260):1443–51.
6. Baladron J, Vitay J, Fietzek T, Id FHH. The contribution of the basal ganglia and cerebellum to motor learning: a neuro-computational approach. PLoS Comput Biol. 2023;19(4):1–29.
7. Li S. Spasticity, motor recovery, and neural plasticity after stroke. Front Neurol. 2017;8(April):1–8.
8. Bologna M, Paparella G, Fasano A, Hallett M, Berardelli A. Evolving concepts on bradykinesia bradykinesia: terminology and clinical aspects. Brain. 2019;2020(143):727–50.
9. Teodoro J, Fernandes S, Castro C, Fernandes JB. Current trends in gait rehabilitation for stroke survivors: a scoping review of randomized controlled trials. J Clin Med. 2024;13(5):1358.
10. Saraiva J, Rosa G, Fernandes S, Fernandes JB. Current trends in balance rehabilitation for stroke survivors: a scoping review of experimental studies. Int J Environ Res Public Health. 2023;20(19):6829–43.
11. Winstein CJ, Stein J, Arena R, Bates B, Cherney LR, Cramer SC, et al. Guidelines for adult stroke rehabilitation and recovery: a guideline for healthcare professionals from the American Heart Association/American Stroke Association. Stroke. 2016;47:e98–e169.
12. Stroke Foundation - Australia. Living stroke guidelines [Internet] 2025 [cited 2025 Sep 24]. Available from: https://strokefoundation.org.au/what-we-do/for-health-professionals/living-stroke-guidelines.

13. Teasell R, Salbach NM, Foley N, Mountain A, Cameron JI, Jong A, et al. Canadian stroke best practice recommendations: rehabilitation, recovery, and community participation following stroke. Int J Stroke. 2020;15(7):763–88.
14. English C, Hillier SL, Lynch EA. Circuit class therapy for improving mobility after stroke. Cochrane Database Syst Rev. 2017;6(6):CD007513.
15. Knox M, Stewart A, Richards CL. Six hours of task-oriented training optimizes walking competency post stroke: a randomized controlled trial in the public health-care system of South Africa. Clin Rehabil. 2018;32(8):1057–68. https://doi.org/10.1002/central/CN-01952135/full.
16. Mehrholz J, Thomas S, Elsner B. Treadmill training and body weight support for walking after stroke. Cochrane Database Syst Rev. 2017;8(8):CD002840.
17. Lerín-Calvo A, González-Carrasco E, Reina-Varona A, Fernández-Pérez JJ. Resistance training for gait rehabilitation in people with stroke. A systematic review and meta-analysis. Disabil Rehabil. 2025;48(2):331–49.
18. Nadeem I, Butt SK, Mubeen I, Razzaq A. Effects of core muscles strengthening exercises with routine physical therapy on trunk balance in stroke patients: a randomized controlled trial. J Pak Med Assoc. 2024;74(5):848–51.
19. Madhuranga PVH, Mathangasinghe Y, Anthony DJ. Improving balance with wobble board exercises in stroke patients: single-blind, randomized clinical trial. Top Stroke Rehabil. 2019;26(8):595–601. https://doi.org/10.1080/10749357.2019.1645439.
20. Gbiri CA, Shittu A. Effects of a six-weeks balance training on balance performance and functional independence in hemiparetic stroke survivors. Indian J Physiother Occup Therapy. 2014;8(4):123.
21. Patten C, Lexell J, Brown HE. Weakness and strength training in persons with poststroke hemiplegia: rationale, method, and efficacy. J Rehabil Res Dev. 2004;41(3A):293–312. Available from: https://pubmed.ncbi.nlm.nih.gov/15543447/.
22. Ada L, Dorsch S, Canning CG. Strengthening interventions increase strength and improve activity after stroke: a systematic review. Australian Journal of Physiotherapy. 2006;52(4):241–8. Available from: https://www.sciencedirect.com/science/article/pii/S0004951406700034?via%3Dihub.
23. Hebert D, Lindsay MP, McIntyre A, Kirton A, Rumney PG, Bagg S, et al. Canadian stroke best practice recommendations: stroke rehabilitation practice guidelines, update 2015. Int J Stroke. 2016;11(4):459–84. Available from: https://pubmed.ncbi.nlm.nih.gov/27079654/.
24. Saunders DH, Sanderson M, Hayes S, Johnson L, Kramer S, Carter DD, et al. Physical fitness training for stroke patients. Cochrane Database Syst Rev. 2020;3(3):1–3. Available from: https://pubmed.ncbi.nlm.nih.gov/32196635/.
25. Mackay-Lyons M, Billinger SA, Eng JJ, Dromerick A, Giacomantonio N, Hafer-Macko C, et al. Aerobic exercise recommendations to optimize best practices in care after stroke: AEROBICS 2019 update. Phys Ther. 2019;100(1):149. Available from https://pmc.ncbi.nlm.nih.gov/articles/PMC8204880/.
26. Saunders DH, Sanderson M, Hayes S, Kilrane M, Greig CA, Brazzelli M, et al. Physical fitness training for stroke patients. Cochrane Database Syst Rev. 2016;3(3). Available from https://pubmed.ncbi.nlm.nih.gov/27010219/.
27. Wang X, Qiu J, Zhou Y, Liu W, Zhang S, Gong Y, et al. Effects of virtual reality-assisted and overground gait adaptation training on balance and walking ability in stroke patients: a randomized controlled trial. Am J Phys Med Rehabil. 2024;103(6):480–7.
28. Akinci M, Burak M, Kasal FZ, Özaslan EA, Huri M, Kurtaran ZA. The effects of combined virtual reality exercises and robot assisted gait training on cognitive functions, daily living activities, and quality of life in high functioning individuals with subacute stroke. Percept Mot Skills. 2024;131(3):756–69.
29. Akıncı M, Burak M, Yaşar E, Kılıç RT. The effects of robot-assisted gait training and virtual reality on balance and gait in stroke survivors: a randomized controlled trial. Gait Posture. 2023;103(July):215–22.
30. Erbil D, Tugba G, Murat TH, Melike A, Merve A, Cagla K, et al. Effects of robot-assisted gait training in chronic stroke patients treated by botulinum toxin-a: a pivotal study. Physiother Res Int. 2018;23(3):1–8.

31. Mehrholz J, Thomas S, Werner C, Kugler J, Pohl M, Elsner B. Electromechanical-assisted training for walking after stroke. Cochrane Database Syst Rev. 2017;5(5):CD006185.
32. Nascimento LR, de Oliveira CQ, Ada L, Michaelsen SM, Teixeira-Salmela LF. Walking training with cueing of cadence improves walking speed and stride length after stroke more than walking training alone: a systematic review. J Physiother. 2015;61(1):10–5.
33. Sharif F. Effectiveness of functional electrical stimulation (FES) versus conventional electrical stimulation in gait rehabilitation of patients with stroke. J Coll Physicians Surg Pak. 2017;27(11):703–6.
34. He W, Yaning L, Shaohong Y. Effect of electrical stimulation in the treatment on patients with foot drop after stroke: a systematic review and network meta-analysis. J Stroke Cerebrovasc Dis. 2025;34(5):108279.
35. Lu W, Shi M, Liu L, Wang S, Deng W, Ma Y, et al. Effect of virtual reality–based therapies on lower limb functional recovery in stroke survivors: systematic review and meta-analysis. J Med Internet Res. 2025;27:e72364.
36. Henrique PPB, Colussi EL, De Marchi ACB. Effects of exergame on patients' balance and upper limb motor function after stroke: a randomized controlled trial. J Stroke Cerebrovasc Dis. 2019;28(8):2351–7.
37. Aphiphaksakul P, Siriphorn A. Home-based exercise using balance disc and smartphone inclinometer application improves balance and activity of daily living in individuals with stroke: a randomized controlled trial. PLoS One. 2022;17(11):1–13. https://doi.org/10.1371/journal.pone.0277870.
38. Bodur BE, Erdoğanoğlu Y, Asena SS. Effects of robotic-assisted gait training on physical capacity, and quality of life among chronic stroke patients: a randomized controlled study. J Clin Neurosci. 2024;120:129–37.
39. Mahmood W, Ahmed Burq HSI, Ehsan S, Sagheer B, Mahmood T. Effect of core stabilization exercises in addition to conventional therapy in improving trunk mobility, function, ambulation and quality of life in stroke patients: a randomized controlled trial. BMC Sports Sci Med Rehabil. 2022;14(1):62.
40. Kumar TMSN, Kumar JS, Rekha K. Comparing the effect of proprioceptive neuromuscular facilitation (PNF) exercise and focused regimen program to assess balance and quality of life for acute stroke rehabilitation. Indian J Physiother Occup Ther. 2024;18:794–9.
41. Yadav R, Walia S, Vats M, Noohu M. Comparison of effectiveness of specific balance strategy training programme with general balance training programme on balance performance in chronic stroke. J Phys Med Rehabil. 2016;19(February):1–6.
42. Gamawa AU, Lawal IU, Salihu AT, Bello UM, Chutiyami M, Salihu D. Effectiveness of an integrated education for self-management in stroke rehabilitation in north-East Nigeria: a randomized controlled trial. Pak J Med Res. 2021;60(1):12–9.
43. Moore SA, Boyne P, Fulk G, Verheyden G, Fini NA. Walk the talk: current evidence for walking recovery after stroke, future pathways and a mission for research and clinical practice. Stroke. 2022;53(11):3494–505. Available from: https://pubmed.ncbi.nlm.nih.gov/36069185/.
44. Fleet JL, McIntyre A, Janzen S, Saikaley M, Qaqish M, Cianfarani R, et al. A systematic review examining the effect of vitamin D supplementation on functional outcomes post-stroke. Clin Rehabil. 2023;37(11):1451–66.
45. Brunner IC, Hansen GM. High-intensity gait training for patients after stroke: a feasibility study. Physiother Res Int. 2025;30(2). Available from: https://pubmed.ncbi.nlm.nih.gov/40207841/.
46. Zhang L, Ma J, Liu X, Jin A, Wang K, Yin X. Cognitive-motor dual-task training on gait and balance in stroke patients: meta-analytic report and trial sequential analysis of randomized clinical trials. J Neuroeng Rehabil. 2024;21(1):227. Available from: https://pubmed.ncbi.nlm.nih.gov/39716165/.
47. Daryabor A, Aminian G, Arazpour M, Baniasad M, Yamamoto S. The effects of ankle-foot orthoses with plantar flexion stop and plantar flexion resistance using rocker-sole shoes on stroke gait: a randomized-controlled trial. Turk J Phys Med Rehabil. 2021;67(4):449–61.

48. Alashram AR, Annino G, Aldajah S, Raju M, Padua E. Rehabilitation of limb apraxia in patients following stroke: a systematic review. Appl Neuropsychol Adult. 2022;29(6):1658–68.
49. Veerbeek JM, Van Wegen E, Van Peppen R, Van Der Wees PJ, Hendriks E, Rietberg M, et al. What is the evidence for physical therapy poststroke? A systematic review and meta-analysis. PLoS One. 2014;9:e87987.
50. Maier M, Ballester BR, Verschure PFMJ. Principles of neurorehabilitation after stroke based on motor learning and brain plasticity mechanisms. Front Syst Neurosci. 2019;13:74.
51. Yu Y, Huang W, Tuerxun H, Zheng Y, Su L, Li X, et al. Enhanced neuroplasticity and gait recovery in stroke patients: a comparative analysis of active and passive robotic training modes. BMC Neurol. 2025;25(1):239.
52. Levack WM, Weatherall M, Hay-Smith EJ, Dean SG, McPherson K, Siegert RJ. Goal setting and strategies to enhance goal pursuit for adults with acquired disability participating in rehabilitation. Cochrane Database Syst Rev. 2015;2015(7):CD009727. https://doi.org/10.1002/14651858.CD009727.pub2. PMID: 26189709; PMCID: PMC8941379..
53. Knutti K, Björklund Carlstedt A, Clasen R, Green D. Impacts of goal setting on engagement and rehabilitation outcomes following acquired brain injury: a systematic review of reviews. Disabil Rehabil. 2022;44(12):2581–90. https://doi.org/10.1080/09638288.2020.1846796. Epub 2020 Nov 19. PMID: 33211986..
54. English C, Hillier SL. Circuit class therapy for improving mobility after stroke. Cochrane Database Syst Rev. 2017;6(6):CD007513.
55. Intercollegiate Stroke Working Party. National clinical guideline for stroke for the UK and Ireland. London: Royal College of Physicians; 2023.
56. Ayelabowo TA, Ajayi TC, Macdonald SHF. Community-based models of mobility training after stroke: a scoping review. Bull Fac Phys Ther. 2024;29(1):61.
57. Thaut MH, Leins AK, Rice RR, Argstatter H, Kenyon GP, McIntosh GC, et al. Rhythmic auditory stimulation improves gait more than NDT/Bobath training in near-ambulatory patients early poststroke: a single-blind, randomized trial. Neurorehabil Neural Repair. 2007;21(5):455–9.
58. An Z, Li K, Yang X, Ke J, Xu Y, Zhang X, et al. Community-based rehabilitation services implemented by multidisciplinary teams among adults with stroke: a scoping review with a focus on Chinese experience. BMC Public Health. 2024;24(740):1–14.

Contextualised Cognitive and Perceptual Rehabilitation Post-Stroke

5

Sureshkumar Kamalakannan, Kamal Narayan Arya, and Sonal V Chitnis

Abstract

This chapter focuses on the rehabilitation of cognitive–perceptual deficits, one of the most important issues among stroke survivors. These impairments multiply post-stroke disability, especially in countries with limited resources. Cognitive deficits range from attention to memory involvement, whereas perceptual impairments vary from spatial neglect to agnosia. These issues affect the functional performance of an individual who needs evidence-based, individualised, and contextualised intervention by multidisciplinary professionals. Various restorative and compensatory rehabilitation techniques may be imparted to lessen the effect of the impairments. Additionally, the importance and implications of various factors affecting evidence-based practice in low-resource settings are also emphasised.

Keywords

Contextualised cognitive rehab · Stroke · LMICs

S. Kamalakannan
Faculty of Health and Life Sciences, Department of Social Work, Education, and Community Wellbeing, Northumbria University, Newcastle upon Tyne, UK
e-mail: sureshkumar.kamalakannan@northumbria.ac.uk

K. N. Arya (✉)
Neuro-Rehab Lab, Department of Occupational Therapy, Pandit Deendayal Upadhyaya National Institute for Persons with Physical Disabilities, New Delhi, India

S. V Chitnis
UNM Foundation, Sankalan Neurorehabilitation Centre, Gujarat, India

PhD Scholar, Bharati Vidyapeeth (Deemed to Be University), School of Audiology & Speech-Language Pathology, Pune, India

D. B. C. Gandhi, J. D. Pandian (eds.), *Stroke Rehabilitation in Low-Resource Settings*, https://doi.org/10.1007/978-981-92-0403-8_5

5.1 Background

Cerebrovascular accident (CVA) or stroke is a major cause of neurological disabilities arising from sensory-motor, cognitive–perceptual, and executive dysfunctions resulting from brain damage [1–4]. These impairments result in various disabilities leading to poor quality of life and independence in everyday functioning [5]. Both cerebral and cerebellar strokes across the lifespan can have a significant impact on everyday life skills, especially cognition, perception, communication, and social skills [6]. The impairments in executive functions and visuospatial skills are associated with more disabling outcomes due to low levels of engagement during stroke care, as these services may be inaccessible, unavailable, or unaffordable in many countries, particularly for individuals living in poverty and rural areas [7]. The disability burden is borne substantially by low- and middle-income countries (LMICs), and the impact of disability is worse among those who are disadvantaged and vulnerable, for example, people living in poverty and in rural areas [8]. This is particularly because of the lack of accessibility to stroke care and rehabilitation in a global situation [9]. People affected by cognitive–perceptual dysfunctions following a stroke may experience much poorer outcomes as therapy services for these areas are not accessible, available, and affordable in many countries [10]. However, understanding cognitive–perceptual dysfunctions following a stroke is of significant clinical relevance in enabling a stroke survivor to engage independently and productively in the community [7].

5.2 Cognitive and Perceptual Dysfunctions Following Stroke

Cognitive dysfunction following a stroke can vary widely depending on the site and extent of the insult to the brain during a stroke [11]. It often includes impairments in attention, concentration, orientation, memory, executive function, communication or language, logical reasoning, and problem-solving abilities [12]. The critical issue for stroke survivors with cognitive deficits is their ability to independently plan, organise, or process information, and their ability to maintain focus on specific tasks and manage new information [11]. These cognitive issues could substantially impact everyday functioning, affecting the ability of the stroke survivor to engage in his or her routine daily living activities such as brushing, grooming, dressing, toileting, childcare, home management, etc. Subsequently, it may impact their ability to fulfil their roles and routines at work and participate effectively in society [12]. Therefore, cognitive rehabilitation strategies, including cognitive therapy and targeted cognitive exercises, can help improve or manage these deficits, but the extent of recovery often depends on the severity of cognitive impairment and the intensity of cognitive rehabilitation [13].

Perceptual dysfunctions post-stroke are a common symptom of stroke. Individuals experiencing perceptual deficits will not be able to accurately interpret sensory information from various sensory modalities in their body [14]. This will lead to difficulties in perceiving objects, space, and their own body. Common problems experienced by a stroke survivor with perceptual dysfunctions include visual neglect, where the stroke survivor will not be able to notice or attend to one side of their visual field. This is common in those with right brain damage, with the left side of the body being affected [15]. Other perceptual problems may involve issues with joint position sense, joint movement sense, spatial deficits such as the inability to interpret foreground from background, inability to perceive depth, colour, etc. [15] These spatial perceptual issues could eventually make it difficult for the stroke survivor to judge the distance between themselves and neighbouring objects or space, which could subsequently limit the stroke survivor's ability to navigate their own environment [16]. Stroke survivors with perceptual dysfunctions may also have trouble in recognising familiar faces and objects termed as agnosia. Agnosia can also impact a stroke survivor's independence in everyday living, such as their ability to dress, groom, eat, and walk safely [17]. Early perceptual rehabilitation and targeted therapies are extremely important for remediating and improving the perceptual skills of the stroke survivor and helping them regain independence [18].

5.3 Cognitive–Perceptual Rehabilitation

Given the impact of post-stroke cognitive and perceptual impairments on the functioning and quality of life, it is important to ensure cognitive–perceptual rehabilitation is accessible to those requiring it, and it must be provided at the earliest opportunity to enable post-stroke independence in a stroke survivor [19]. Evidence-based patient-centred interventions have tremendous therapeutic benefit for regaining independence following a stroke [20]. Although cognitive–perceptual rehabilitation might sound like an intervention that could be provided by professionals related to a single discipline like a neuropsychologist or an occupational therapist, it is complex. It is typically provided by multidisciplinary professionals through a combination of individualised therapeutic interventions structured according to the needs of the affected individual [21]. It is usually delivered by various stroke care professionals including but not limited to a neuropsychologist, occupational therapist, speech-language therapist, rehabilitation nurse, and physiotherapist [21]. The cognitive rehabilitation process starts with a comprehensive assessment that would confirm the specific cognitive–perceptual deficits post -stroke. Based on the findings from the assessment, a comprehensive therapy plan would be developed to address the deficits [22]. There are two main approaches to cognitive–perceptual rehabilitation. The first one is a remedial approach targeted towards remediating the cognitive–perceptual deficits, and the second one is targeted towards training the stroke survivor to compensate for the deficits using adaptive techniques or adaptive equipment with which they can regain independence [23]. The intervention can be delivered in a variety of settings and formats in the hospital as well as in the communities. The focus of any cognitive–perceptual rehabilitation is usually enabling functioning, independence in everyday living, and the overall quality of life of the individual affected [23].

5.4 Approaches to Cognitive–Perceptual Rehabilitation in Stroke Care

Cognitive–perceptual impairment greatly affects the functional performance of a post-stroke survivor [24]. An individual with left hemiparesis may not be able to retrieve items related to functional tasks in his left hemi-space (Fig. 5.1). Another stroke patient may not be able to respond to an object such as a glass kept on a table and for drinking water in spite of having some motor abilities in the paretic upper limb. An array of cognitive–perceptual deficits, such as unilateral spatial neglect, apraxia, and decline in memory and executive functions, may be experienced after the brain insult. The management, including assessment of these impairments, has always been a challenge [25]. The therapeutic interventions range from restorative to compensatory rehabilitation and from general to specific deficits (Table 5.1).

Prism adaptation is a method to provide visuo-motor training by using optical prisms. The optical prism alters the vision towards the right side in an individual with left body paresis, enhancing the limb movement in the left hemi-space. The patient undergoes task practice with the modified vision, usually for 15–20 min. With repetitive training, the quality of functional performance enhances, which, in turn, impacts daily functions such as eating, transfers, and ambulation [34].

Mirror therapy, a simple and economical method of providing the visual illusion for the neglected hemi-space as well as the paretic upper limb, has potential for delineating the impact of neglect in post-stroke individuals. In this technique, the

Fig. 5.1 A post-stroke patient with left hemiparesis is unable to perceive (despite normal vision) the food material on the left side of the utensil while eating with the right upper extremity. The inability is due to unilateral spatial neglect

Table 5.1 Intervention approaches used for various cognitive perceptual deficits in stroke

Impairment	Definition	Approaches	Application in low-resource settings	Evidence (level 1A)
Unilateral spatial neglect	Impairment in perceiving sensory or motor information on the contra-lesional side of the brain damage	(a) Prism adaptation	Yes	Very low-certainty evidence for neglect outcomes and activities of daily living [26]
		(a) Mirror therapy	Yes	Effective for improving activities of daily living [27]
		(b) Cueing/feedback	Yes	*Not available*
		(c) Visual scanning	Yes	Very low-certainty evidence for neglect outcomes and activities of daily living [26]
		(d) Virtual reality	No	Limited evidence [27]
		(e) Robotic training	No	Inconclusive [28]
Attention deficit	Impaired ability to respond, attend, or select a specific stimulus	(a) Restorative cognitive training	Yes	Inconclusive [29]
		(b) Compensatory rehabilitation	Yes	
Memory deficit	Impairment in recalling the experienced incidents (episodic memory)	(a) Restorative cognitive training	Yes	Limited evidence [30]
		(b) Compensatory rehabilitation	Yes	
Apraxia	Inability to execute skilled functional movement that cannot be explained on the basis of motor or sensory deficit	(a) Restorative cognitive training	Yes	Some evidence for the cognitive, motor, and daily functions [31]
		(b) Compensatory rehabilitation	Yes	
Other generic/associated cognitive rehabilitation approaches				
Physical exercises			Yes	Small to moderate evidence for the cognitive functions [32]
Dual-task training			Yes	Limited evidence [33]
Non-invasive brain stimulation			No	Very low-certainty evidence for activities of daily living [26]

patient is in a seated position near a table on which a mirror box has been placed, opposite to the midline of the body. The patient places his paretic upper limb inside the box, whereas the less-paretic limb is opposite to the mirror. The movement of the less-paretic limb is being performed, preferably through object-based activities such as squeezing a ball and lifting pegs. The movements are perceived as illusionary movement of the paretic limb [27].

The *cueing/feedback* technique comprises providing visual, verbal, or sensory cues during the activities demanding interaction with information in the neglected hemi-space. Paper–pencil tasks such as letter or star cancellations, reading, writing, and drawing can be utilised for this approach. Cueing in the form of highlighting with different colours, focal light, and providing the items or feedback from the neglected hemi-space side are some of the ways to impart the technique [35]. Similarly, the *visual scanning* method focuses on enhancing scanning behaviour through tasks such as digit detection on computer screens, picture exploration, and copying, reading, and writing tasks. Repetitiveness, error correction, feedback, and reinforcement are also provided for the training [36].

Virtual reality is a method in which a post-stroke patient interacts with the computerised environment in immersive, semi-immersive, or non-immersive form. Various computer-formulated daily functions and games are used to impart the training. The method allows the individual to interact with the virtual environment, enhancing awareness of the neglected space. Such types of interaction have been found to be interesting and adhering for the patient [37].

Robotic training provides highly intensive, adaptive, and task-specific training in passive or active-assistive mode. The intervention enhances the midline perception among stroke patients with neglect [29]. The availability and cost of virtual reality and robotic training are always a challenge in clinical practice.

Restorative cognitive training comprises of repetitive practice of activities having specific cognitive demands. The practice leads to the learning of cognitive skills by strengthening the weakened behaviour [38]. The change is assumed to occur by cortical reorganisation. Activities such as puzzles, table-top games, paper–pencil tasks, and peg boards (variety of shapes and colours) are utilised for the training. Fig. 5.2 shows the pre- and post-intervention (writing activity) changes in the severity of spatial neglect.

Compensatory rehabilitation is an approach based on compensating for the impaired cognition using specific techniques, training, devices, or adaptations with the objective of independently performing cognitively related daily functions. Some of the compensation methods are checklists, the use of digital timers or reminders, adapted calendars or planners, time scheduling, mnemonics, and environmental modification. These simple methods are supportive in memory and attention rehabilitation [30, 31]. The repetitive practice of functional tasks leading to enhanced skill in that particular task is also utilised in compensatory rehabilitation, though the favourable change may also be attributed to the restorative aspect of cognitive rehabilitation.

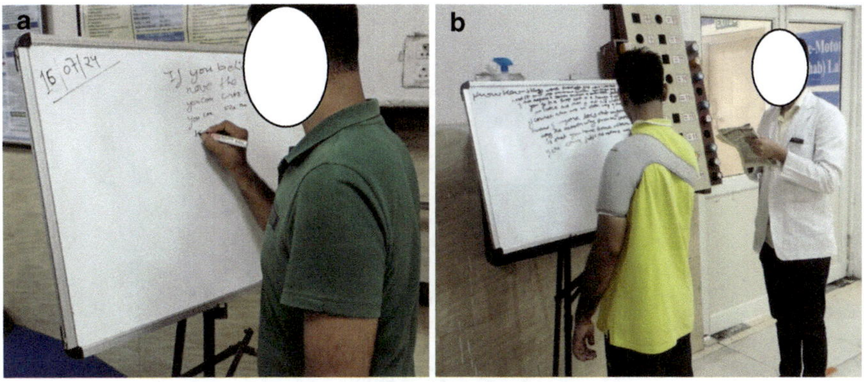

Fig. 5.2 (**a**, **b**) A post-stroke patient with left hemiparesis and unilateral spatial neglect is performing a writing activity (by the right hand) on a whiteboard. Initially, he was unable to write on the left side of the board (**a**). After 12 weeks of training with the same activity, he is able to expand his writing on the left side of the board (**b**)

In addition to the above approaches, in low-resource settings, structured physical activities may be considered as an easy, beneficial, and economically viable solution for restoring overall cognitive abilities among stroke survivors [33]. Exercises in the form of resistive exercise for the paretic as well as non-paretic side, walking, stair climbing, and sit-to-stand may be provided with the dual objectives of fitness and enhanced cognition. Physical activities or exercises, along with dual-task demanding cognitive abilities, are also valuable in enhancing the cognitive function among stroke patients [39]. Dual tasks may be considered in the form of cognitive–motor activities; for instance, reverse counting during forward walking and vice versa. Non-invasive brain stimulation (NIBS) such as transcranial direct current stimulation may also be combined with cognitive training to modulate neural reorganisation, augmenting cognitive function [40]. Physical exercises and dual-task training have more application in low-resource settings, whereas NIBS has limited application and a lack of strong evidence.

5.5 Contextual Barriers to Cognitive–Perceptual Rehabilitation

The context in which one experiences a stroke and cognitive–perceptual deficit plays a crucial role in stroke rehabilitation [41]. The demarcating reason for this is the resources and systems available in those contexts to provide good quality, safe, and effective cognitive–perceptual rehabilitation following a stroke. Rehabilitation of a stroke survivor with cognitive–perceptual dysfunction in a high-income country, for example, the United Kingdom, is significantly different from that in a low- and middle-income country (LMIC) like India [42]. There are several other factors

Fig. 5.3 Factors influencing contextualised stroke care

that would define a context, such as language, culture, values, and beliefs. And these contextual factors could sometimes act as an advantage and sometimes a disadvantage. For example, in India, support from immediate family members of the stroke survivor enhances rehabilitation and recovery throughout the continuum of care. However, given the lack of organised stroke care even in tertiary hospitals, despite the family support, inaccessibility to stroke rehabilitation services could be a major disadvantage [43]. Figure 5.3 describes the factors influencing contextualised stroke care.

Some of the key contextual factors that could impact cognitive–perceptual rehabilitation following stroke are:

1. Availability of professionals involved in the provision of cognitive–perceptual rehabilitation.
2. Affordability of the stroke survivor and their families for the comprehensive services they may need.
3. Accessible systems and pathways in place to ensure the gains from therapy are translated into regaining function and independence.

4. Approach (remedial versus compensatory) that the stroke care experts take to support the stroke survivor regain function and independence in everyday life.
5. Acceptability of the therapeutic support and care received by the stroke survivor and the experience of the family members in stroke care.

5.6 Strategies to Promote Contextualised Cognitive–Perceptual Rehabilitation

It is important to consider the above-mentioned broad barriers that influence the process and outcomes of rehabilitation after stroke, to contextualise cognitive–perceptual rehabilitation (). It is also important to remember that the contextualisation depends on two other important factors related to the severity of stroke and the evidence to inform decision making and caring for someone with cognitive–perceptual dysfunctions [44]. Given the range of factors that influence contextualisation, it is important to understand the strategies that could be optimised for contextualisation of cognitive–perceptual rehabilitation in stroke care.

5.6.1 Incorporate Felt Needs and Family Situation in the Care Plan

This involves understanding the needs of the stroke survivor requiring cognitive–perceptual rehabilitation and the family situation related to caring and their financial status. As this is crucial to align the goals of the rehabilitation with the needs of the context in which one experiences stroke and make it more realistic and achievable [45]. The intervention will also be sensitive to the culture, values, and beliefs of the stroke survivor and their family.

5.6.2 Contextualise the Evidence

Although cognitive–perceptual rehabilitation is extensively evidence-based, much of this evidence comes from high-resourced settings, which may not fit the purpose in low-resource settings where much of the stroke burden exists [46]. Therefore, systematically develop and evaluate interventions that suit the purpose, but more importantly the context.

5.6.3 Build the Capacity for Organised Provision of Stroke Care

It is important that cognitive–perceptual rehabilitation available to the stroke survivor is safe, comprehensive, multidisciplinary, and with optimal intensity [45]. This is only possible when professionals connect to each other and work together as a team to enhance the outcomes of cognitive–perceptual rehabilitation and also translate the efforts at the hospital or rehabilitation centre from available professionals into sustainable benefits for the stroke survivor in the community [46].

5.6.4 Ensure Continuum of Care in the Community Using Technology

Integration of technology to connect stroke experts at hospitals or rehabilitation centres to stroke survivors requiring cognitive–perceptual rehabilitation in their own communities or home is of utmost importance [43]. This is especially because it is not possible to retain the functional and cognitive–perceptual gains obtained during a stroke survivor's stay in hospital or rehabilitation centre in their own environment, as they cannot be followed up effectively if the cognitive–perceptual rehabilitation focuses on the impairments rather than the context [45]. Hence, this is crucial, and this gap in the continuum of care in the community can be bridged by optimising technology, connecting the stroke survivor in the community with the experts at hospitals, and ensuring seamless rehabilitation [43]. However, technological solutions may also imply that these interventions depend on aspects such as digital literacy and access to devices, which can vary greatly among individuals and the contexts.

Ultimately, cognitive–perceptual rehabilitation is best contextualised by recognising various factors as described above with an evidence-based approach to ensure the rehabilitation delivered and experiences are safe, effective, and holistic, tailored to the felt needs of the stroke survivor.

5.7 Implications of Contextualised Cognitive–Perceptual Rehabilitation

There are several important implications of contextualised cognitive–perceptual rehabilitation for research, practice, and policy.

5.8 Practice Implications

The expertise required to provide cognitive–perceptual rehabilitation is evolving as the practice of diverse techniques informs research and the need to hone certain specific competencies. However, to contextualise stroke care, the three strategies mentioned above are crucial. Practice particularly in low-resource settings cannot apply the evidence coming from high-income countries. The contexts are so different. Similarly, every context in a high-income setting does not function in the same way. Depending on the situation of the providers, the wider systems, and the socio-political support, the need for cognitive–perceptual rehabilitation would change [45]. Hence, practice must be sensitive to the context in which stroke rehabilitation for cognition and perception is provided.

5.9 Policy Implications

There is a dire need to include rehabilitation within the existing policies for stroke in many countries. Available stroke services are predominantly medical, provided in acute settings and the concept of stroke care for stroke survivors in the chronic stage and those in the community are often neglected in the existing policies. In some contexts, even in high-income countries, the waiting list to access stroke rehabilitation, especially cognitive–perceptual rehabilitation services, is very long [47].

5.10 Research Implications

There is a dearth of evidence on contextualised cognitive–perceptual rehabilitation. Existing evidence is targeted more towards improving impairments in cognitive–perceptual dysfunctions following stroke. However, there is an immense need to understand the impact of various contextual factors on cognitive–perceptual rehabilitation. This is much needed for countries with low-resource settings worldwide [48].

5.11 Summary

This chapter provides a comprehensive understanding of contextualised cognitive–perceptual rehabilitation. It describes various cognitive and perceptual dysfunctions following a stroke and the impact such kind of cognitive–perceptual deficits could have on the function and independence of a stroke survivor experiencing such disabilities. It also highlights the various factors that could affect the implementation of contextualised cognitive–perceptual rehabilitation and the implications of it for practice, policy, and research in stroke rehabilitation worldwide.

What Does Evidence Say

- Stroke survivors with attention and concentration deficits may benefit from cognitive rehabilitation, along with structured exercise programs and engagement in leisure activities.
- Individuals with memory impairments may undergo cognitive rehabilitation to support short-term improvement. Strategies can include internal

(continued)

techniques (e.g. association, mental rehearsal, rhymes) and external aids (e.g. diaries, calendars, alarms, audio recordings, photographs, and mobile devices).
- For those with general cognitive impairment, meta-cognitive strategies and structured cognitive training may be provided.
- Stroke survivors with limb apraxia may be treated using approaches such as gesture training, strategy training, and errorless learning.
- Individuals experiencing unilateral neglect may receive cognitive rehabilitation interventions, including computer-based scanning training, pen-and-paper tasks, visual scanning exercises, eye patching, and mental practice.
- Mirror therapy may also be used in cases of unilateral neglect to improve upper limb function and performance in activities of daily living.
- Non-invasive brain stimulation is not recommended for routine clinical use in managing unilateral neglect, but may be considered within a research setting.

References

1. Hazelton C, Thomson K, Todhunter-Brown A, Campbell P, Chung CS, Dorris L, Gillespie DC, Hunter SM, McGill K, Nicolson DJ, Williams LJ, Brady MC. Interventions for perceptual disorders following stroke. Cochrane Database Syst Rev. 2022;11(11):CD007039. https://doi.org/10.1002/14651858.CD007039.pub3.
2. Li X, He Y, Wang D, Rezaei MJ. Stroke rehabilitation: from diagnosis to therapy. Front Neurol. 2024;15:1402729. https://doi.org/10.3389/fneur.2024.1402729.
3. Hazelton C, McGill K, Campbell P, Todhunter-Brown A, Thomson K, Nicolson DJ, Cheyne JD, Chung C, Dorris L, Gillespie DC, Hunter SM. Perceptual disorders after stroke: a scoping review of interventions. Stroke. 2022;53(5):1772–87. https://doi.org/10.1161/STROKEAHA.121.035671.
4. Small R, Wilson PH, Wong D, Rogers JM. Who, what, when, where, why, and how: a systematic review of the quality of post-stroke cognitive rehabilitation protocols. Ann Phys Rehabil Med. 2022;65(5):101623. https://doi.org/10.1016/j.rehab.2021.101623.
5. Cumming TB, Marshall RS, Lazar RM. Stroke, cognitive deficits, and rehabilitation: still an incomplete picture. Int J Stroke. 2013;8(1):38–45. https://doi.org/10.1111/j.1747-4949.2012.00972.x.
6. Gillen G. Cognitive and perceptual rehabilitation: optimizing function. Amsterdam: Elsevier Health Sciences; 2008. https://doi.org/10.1016/B978-0-323-04621-3.X1000-5.
7. Kumar KS, Samuelkamaleshkumar S, Viswanathan A, Macaden AS. Cognitive rehabilitation for adults with traumatic brain injury to improve occupational outcomes. Cochrane Database Syst Rev. 2017;6(6):CD007935. https://doi.org/10.1002/14651858.CD007935.pub2.
8. Pandian JD, Sebastian IA. Integrated approach to stroke burden: are we doing enough? Lancet Neurol. 2021;20(10):774–5. https://doi.org/10.1016/S1474-4422(21)00287-8.
9. GBD 2019 Stroke Collaborators. Global, regional, and national burden of stroke and its risk factors, 1990-2019: a systematic analysis for the global burden of disease study 2019. Lancet Neurol. 2021;20(10):795–820. https://doi.org/10.1016/S1474-4422(21)00252-0.
10. GBD 2017 Cirrhosis Collaborators. The global, regional, and national burden of cirrhosis by cause in 195 countries and territories, 1990-2017: a systematic analysis for the global burden of

disease study 2017. Lancet Gastroenterol Hepatol. 2020;5(3):245–66. https://doi.org/10.1016/S2468-1253(19)30349-8.

11. Elendu C, Amaechi DC, Elendu TC, Ibhiedu JO, Egbunu EO, Ndam AR, Ogala F, Ologunde T, Peterson JC, Boluwatife AI, Okongko AO, Fatoye JO, Akpovona OL, Onyekweli SO, Temitope AY, Achimugu AO, Temilade AV. Stroke and cognitive impairment: understanding the connection and managing symptoms. Ann Med Surg. 2023;85(12):6057–66. https://doi.org/10.1097/MS9.0000000000001441.
12. Donovan NJ, Kendall DL, Heaton SC, Kwon S, Velozo CA, Duncan PW. Conceptualizing functional cognition in stroke. Neurorehabil Neural Repair. 2008;22(2):122–35. https://doi.org/10.1177/1545968307306239.
13. Gray M, Gills JL, Glenn JM, Vincenzo JL, Walter CS, Madero EN, Hall A, Fuseya N, Bott NT. Cognitive decline negatively impacts physical function. Exp Gerontol. 2021;143:111164. https://doi.org/10.1016/j.exger.2020.111164.
14. Hazelton C, Todhunter-Brown A, Campbell P, Thomson K, Nicolson DJ, McGill K, Chung CS, Dorris L, Gillespie DC, Hunter SM, Williams LJ, Brady MC. Interventions for people with perceptual disorders after stroke: the PIONEER scoping review, Cochrane systematic review and priority setting project. Health Technol Assess. 2024;28(69):1–141. https://doi.org/10.3310/WGJT3471.
15. Gorst T, Rogers A, Morrison SC, Cramp M, Paton J, Freeman J, Marsden J. The prevalence, distribution, and functional importance of lower limb somatosensory impairments in chronic stroke survivors: a cross sectional observational study. Disabil Rehabil. 2019;41(20):2443–50. https://doi.org/10.1080/09638288.2018.1468932.
16. Vancleef K, Colwell MJ, Hewitt O, Demeyere N. Current practice and challenges in screening for visual perception deficits after stroke: a qualitative study. Disabil Rehabil. 2022;44(10):2063–72. https://doi.org/10.1080/09638288.2020.1824245.
17. Meyer S, Karttunen AH, Thijs V, Feys H, Verheyden G. How do somatosensory deficits in the arm and hand relate to upper limb impairment, activity, and participation problems after stroke? A systematic review. Phys Ther. 2014;94(9):1220–31. https://doi.org/10.2522/ptj.20130271.
18. Bowen A, Knapp P, Gillespie D, Nicolson DJ, Vail A. Non-pharmacological interventions for perceptual disorders following stroke and other adult-acquired, non-progressive brain injury. Cochrane Database Syst Rev. 2011;2011(4):CD007039. https://doi.org/10.1002/14651858.CD007039.pub2.
19. Cook EA, Thigpen R. Identification and management of cognitive and perceptual deficits in the rehabilitation patient. Rehabil Nurs. 1993;18(5):310–3. https://doi.org/10.1002/j.2048-7940.1993.tb00777.x.
20. Tyson SF, Burton L, McGovern A. Multi-disciplinary team meetings in stroke rehabilitation: an observation study and conceptual framework. Clin Rehabil. 2014;28(12):1237–47. https://doi.org/10.1177/0269215514535942.
21. Kanevsky D, Yabrov A. Aspects of perceptual development and rehabilitation. Med Hypotheses. 1988;27(1):33–4. https://doi.org/10.1016/0306-9877(88)90079-5.
22. Mulhern M. Cognitive rehabilitation interventions for post-stroke populations. Dela J Public Health. 2023;9(3):70–4. https://doi.org/10.32481/djph.2023.08.012.
23. Gibson E, Koh CL, Eames S, Bennett S, Scott AM, Hoffmann TC. Occupational therapy for cognitive impairment in stroke patients. Cochrane Database Syst Rev. 2022;3:CD006430.
24. Cramer SC, Richards LG, Bernhardt J, Duncan P. Cognitive deficits after stroke. Stroke. 2023;54:5–9.
25. Longley V, Hazelton C, Heal C, Pollock A, Woodward-Nutt K, Mitchell C, Pobric G, Vail A, Bowen A. Non-pharmacological interventions for spatial neglect or inattention following stroke and other non-progressive brain injury. Cochrane Database Syst Rev. 2021;7:CD003586.
26. Zhang Y, Xing Y, Li C, Hua Y, Hu J, Wang Y, Ya R, Meng Q, Bai Y. Mirror therapy for unilateral neglect after stroke: a systematic review. Eur J Neurol. 2022;29:358–71.
27. Ogourtsova T, Souza Silva W, Archambault PS, Lamontagne A. Virtual reality treatment and assessments for post-stroke unilateral spatial neglect: a systematic literature review. Neuropsychol Rehabil. 2017;27:409–54.

28. Bazan R, Fonseca BHS, Miranda JMA, Nunes HRC, Bazan SGZ, Luvizutto GJ. Effect of robot-assisted training on unilateral spatial neglect after stroke: systematic review and meta-analysis of randomized controlled trials. Neurorehabil Neural Repair. 2022;36:545–56.
29. Loetscher T, Potter KJ, Wong D, das Nair R. Cognitive rehabilitation for attention deficits following stroke. Cochrane Database Syst Rev. 2019;2019:CD002842.
30. da Nair R, Cogger H, Worthington E, Lincoln NB. Cognitive rehabilitation for memory deficits after stroke. Cochrane Database Syst Rev. 2016;9:CD002293.
31. Alashram AR, Annino G, Aldajah S, Raju M, Padua E. Rehabilitation of limb apraxia in patients following stroke: a systematic review. Appl Neuropsychol Adult. 2022;29:1658–68.
32. Oberlin LE, Waiwood AM, Cumming TB, Marsland AL, Bernhardt J, Erickson KI. Effects of physical activity on poststroke cognitive function: a meta-analysis of randomized controlled trials. Stroke. 2017;48:3093–100.
33. He Y, Yang L, Zhou J, Yao L, Pang MYC. Dual-task training effects on motor and cognitive functional abilities in individuals with stroke: a systematic review. Clin Rehabil. 2018;32:865–77.
34. Gammeri R, Schintu S, Salatino A, Vigna F, Mazza A, Gindri P, Barba S, Ricci R. Effects of prism adaptation and visual scanning training on perceptual and response bias in unilateral spatial neglect. Neuropsychol Rehabil. 2024;34:155–80.
35. Turgut N, Möller L, Dengler K, Steinberg K, Sprenger A, Eling P, Kastrup A, Hildebrandt H. Adaptive cueing treatment of neglect in stroke patients leads to improvements in activities of daily living: a randomized controlled, crossover trial. Neurorehabil Neural Repair. 2018;32:988–98.
36. van Kessel ME, Geurts AC, Brouwer WH, Fasotti L. Visual scanning training for neglect after stroke with and without a computerized lane tracking dual task. Front Hum Neurosci. 2013;7:358.
37. Salatino A, Zavattaro C, Gammeri R, Cirillo E, Piatti ML, Pyasik M, Serra H, Pia L, Geminiani G, Ricci R. Virtual reality rehabilitation for unilateral spatial neglect: a systematic review of immersive, semi-immersive and non-immersive techniques. Neurosci Biobehav Rev. 2023;152:105248.
38. Jeffares I, Merriman NA, Rohde D, McLoughlin A, Scally B, Doyle F, Horgan F, Hickey A. A systematic review and meta-analysis of the effects of cardiac rehabilitation interventions on cognitive impairment following stroke. Disabil Rehabil. 2021;43(6):773–88. https://doi.org/10.1080/09638288.2019.1641850.
39. Feng QM, Chen JW. Effects of exercise combined with cognitive dual-task training on cognitive function and sustained effects in stroke patients: a systematic review and meta-analysis. Disabil Rehabil. 2025;47(14):3538–47.
40. Draaisma LR, Wessel MJ, Hummel FC. Non-invasive brain stimulation to enhance cognitive rehabilitation after stroke. Neurosci Lett. 2020;719:133678.
41. Dawar D, Kamalakannan S, Chawla NS, Mathew JT, Mehmood E, Bhatnagar U, Pandian JD. A scoping review of recent advancements in intervention and outcome measures for post-stroke cognitive impairments. Ann Indian Acad Neurol. 2023;26(Suppl 1):S32–42. https://doi.org/10.4103/aian.aian_1099_21.
42. Gillespie DC, Bowen A, Chung CS, Cockburn J, Knapp P, Pollock A. Rehabilitation for post-stroke cognitive impairment: an overview of recommendations arising from systematic reviews of current evidence. Clin Rehabil. 2015;29(2):120–8. https://doi.org/10.1177/0269215514538982.
43. Kamalakannan S, Gudlavalleti Venkata M, Prost A, Natarajan S, Pant H, Chitalurri N, Goenka S, Kuper H. Rehabilitation needs of stroke survivors after discharge from hospital in India. Arch Phys Med Rehabil. 2016;97(9):1526–32. https://doi.org/10.1016/j.apmr.2016.02.008.

44. Murphy CP, Jackson RC, Cooke K, Roca A, Benguigui N, Williams AM. Contextual information and perceptual-cognitive expertise in a dynamic, temporally-constrained task. J Exp Psychol Appl. 2016;22(4):455–70. https://doi.org/10.1037/xap0000094.
45. Sidek NN, Kamalakannan S, Tengku Ismail TA, Musa KI, Ibrahim KA, Abdul Aziz Z, Papachristou NI. Experiences and needs of the caregivers of stroke survivors in Malaysia-a phenomenological exploration. Front Neurol. 2022;13:996620. https://doi.org/10.3389/fneur.2022.996620.
46. Gandhi DBC, Kamalakannan S, Urimubenshi G, Sebastian IA, Montanaro VVA, Chawla NS, D'souza JV, Ngeh E, Mahmood A, Demers M, Hombali A, Solomon JM, Collaborators GCSR. Stroke rehabilitation clinical practice guidelines in low- and middle-income countries: a systematic review of quality and unique features. Cerebrovasc Dis. 2024; https://doi.org/10.1159/000539999.
47. Ghaffar A, Bachani AM, Hyder AA, Cieza A, Bhangu A, Bussières A, Sanchez-Ramirez DC, Gandhi DBC, Verbunt J, Rasanathan K, Gustafsson L, Côté P, Reebye R, De la Cerna-Luna R, Negrini S, Frontera WR, Kamalakannan S. Role and promise of health policy and systems research in integrating rehabilitation into the health systems. Health Res Policy Syst. 2024;22(1):143. https://doi.org/10.1186/s12961-024-01235-2.
48. Kamalakannan S, Srivastava A, Chockalingam M, Gandhi D, Dhamija RK, Solomon J, Akkunje PS, Chitnis S, Someshwar H, Surya N. Health systems for rehabilitation - critical appraisal of the National Programme for stroke in India. J Indian Fed Neurorehabilit. 2024;1(1):18–26.

Contextualized Speech–Language and Swallow Rehabilitation Post Stroke in Low-Resource Settings: Strategies for Aphasia, Dysarthria, Apraxia of Speech, Right Hemisphere Language Disorder, and Dysphagia

6

Preetie Shetty Akkunje, Thejaswi Dodderi, and Sonal V Chitnis

Abstract

Stroke is a leading cause of communication and swallowing impairments globally, disproportionately affecting individuals in low- and middle-income countries (LMICs). Despite the rising prevalence of stroke-related disabilities, access to speech–language–swallowing rehabilitation remains limited due to multidimensional systemic, cultural, and resource-related constraints. This chapter explores context-specific challenges and evidence-based interventions for managing neurogenic communication disorders such as aphasia, dysarthria, apraxia of speech, right hemisphere language disorder, and neurogenic dysphagia in LMICs. Barriers such as shortages of trained speech–language pathologists (SLPs), limited access to cutting-edge diagnostic and therapeutic tools, cultural–

P. S. Akkunje (✉)
Department of Audiology and Speech Language Pathology, Kasturba Medical College Mangalore, Manipal Academy of Higher Education, Manipal, India

Music and Health Science Research Collaboratory (MAHRC), University of Toronto, Toronto, ON, Canada
e-mail: preetie.shetty@manipal.edu

T. Dodderi
Nitte (Deemed to be University), Nitte Institute of Speech and Hearing (NISH), Mangalore, India

S. V Chitnis
UNM Foundation, Sankalan Neurorehabilitation Centre, Gujarat, India

PhD Scholar, Bharati Vidyapeeth (Deemed to Be University), School of Audiology & Speech-Language Pathology, Pune, India

D. B. C. Gandhi, J. D Pandian (eds.), *Stroke Rehabilitation in Low-Resource Settings*, https://doi.org/10.1007/978-981-92-0403-8_6

linguistic diversity, financial constraints, and the lack of governmental policy support for the integration of SLP services into stroke rehabilitation programs are discussed. To constructively address these challenges in LMICs, the chapter sincerely puts forth evidence-based pragmatic rehabilitation approaches based on principles of neural plasticity, motor learning, and exercise physiology for neurogenic communication disorders and neurogenic dysphagia. The chapter also discusses low-cost, culturally adapted interventions such as communication partner training, group therapy, task shifting to community health workers, and the use of regionally available materials in LMICs. The untapped potential of telerehabilitation as a scalable solution for SLPs is also highlighted in the chapter. Lastly, the chapter emphasizes the need for innovative, sustainable, and culturally responsive solutions to improve communication and swallowing outcomes in stroke survivors, ultimately enhancing their quality of life and societal reintegration in LMICs.

Keywords

Stroke rehabilitation · Speech–language pathology · Neurogenic communication disorder · Neurogenic dysphagia · Good health and well-being

Learning Objectives
By the end of the chapter, readers must be able to:

1. Understand the impact of stroke on neurogenic communication disorders and neurogenic dysphagia from the International Classification of Functioning, Disability and Health framework.
2. Identify the barriers and challenges hindering the application of evidence-based speech–language–swallow rehabilitation in low-middle-income countries.
3. Know what and how to implement effective, culturally adapted evidence-based rehabilitation solutions in low-resource settings.
4. Disseminate the knowledge in their clinical practice and proactively advocate for improved stroke rehabilitation services in their workplace.

6.1 Introduction

Globally, stroke is the third most common cause of neurological disabilities, with stroke survivors in low- and middle-income countries (LMICs) experiencing higher morbidity, mortality, and significant functional limitations [1]. Cerebral and cerebellar strokes in adults commonly affect speech production [2], language [3], reading and writing [4], cognitive-communication [5], and swallowing functions [6], thereby restricting employment, social participation and overall quality of life

(QoL) [7, 8]. Figure 6.1 illustrates post-stroke speech–language–swallowing impairments using the International Classification of Functioning, Disability, and Health framework. The cyclic arrows depict the interplay between body structure and function, activity limitation, and participation restrictions. The immediate effects and long-term outcomes in Fig. 6.1 highlight the pressing need for *'early identification and intervention'* post-stroke. Over time, unattended post-stroke conditions can lead to poor functional outcomes, including impaired communication [3], increased caregiver burden [9], psychological distress [10], and reduced overall QoL [11]. Hence, early and systematically graded speech–language–swallowing therapy, by qualified speech–language pathologists (SLPs), is critical for optimizing neuroplasticity and functional recovery, thereby improving post-stroke outcomes and restoring premorbid levels of communication and swallowing function [9, 10].

Speech–language–swallowing therapy has been reasonably established in countries like India [11, 12], Egypt [13], and the Philippines [14]. However, across LMICs, there is a scarcity of trained SLPs for stroke neurorehabilitation [15, 16], with certain regions having rapid growth in service delivery, driven by increased public awareness and a growing emphasis on rehabilitation within healthcare systems [17]. Additionally, in LMICs, accessibility, affordability, and cultural–linguistic diversity pose significant challenges to implementing ecologically valid outcome measures and therapeutic protocols, primarily designed for English-speaking populations [18]. Stigma and culturally rooted beliefs regarding disability may adversely influence help-seeking behaviors post-stroke. Limited access to technological advancements, such as videofluoroscopic swallowing studies [19] and assistive communication devices [20], necessitates the development of low-cost, context-specific solutions by SLPs. This chapter highlights the importance of tailoring interventions to regional needs, leveraging available resources, and promoting inclusive rehabilitation practices within diverse healthcare settings.

6.2 Rehabilitation Approaches for Post-Stroke Speech–Language–Swallowing Disorders

Evidence-based stroke rehabilitation is grounded in three core principles: (a) neural plasticity [21], (b) motor learning [22], and (c) exercise physiology [23]. Neural plasticity is the brain's capacity to reorganize and form new connections [24]. Motor learning emphasizes skill acquisition through repetition and feedback [25], while exercise physiology enhances muscle strength and endurance for functional improvements [23]. Together, these principles guide the development of targeted, adaptive rehabilitation protocols that promote experience-dependent recovery post-stroke. The following section outlines the feasibilities, best practices, and evidence-based recommendations for managing neurogenic communication disorders and dysphagia.

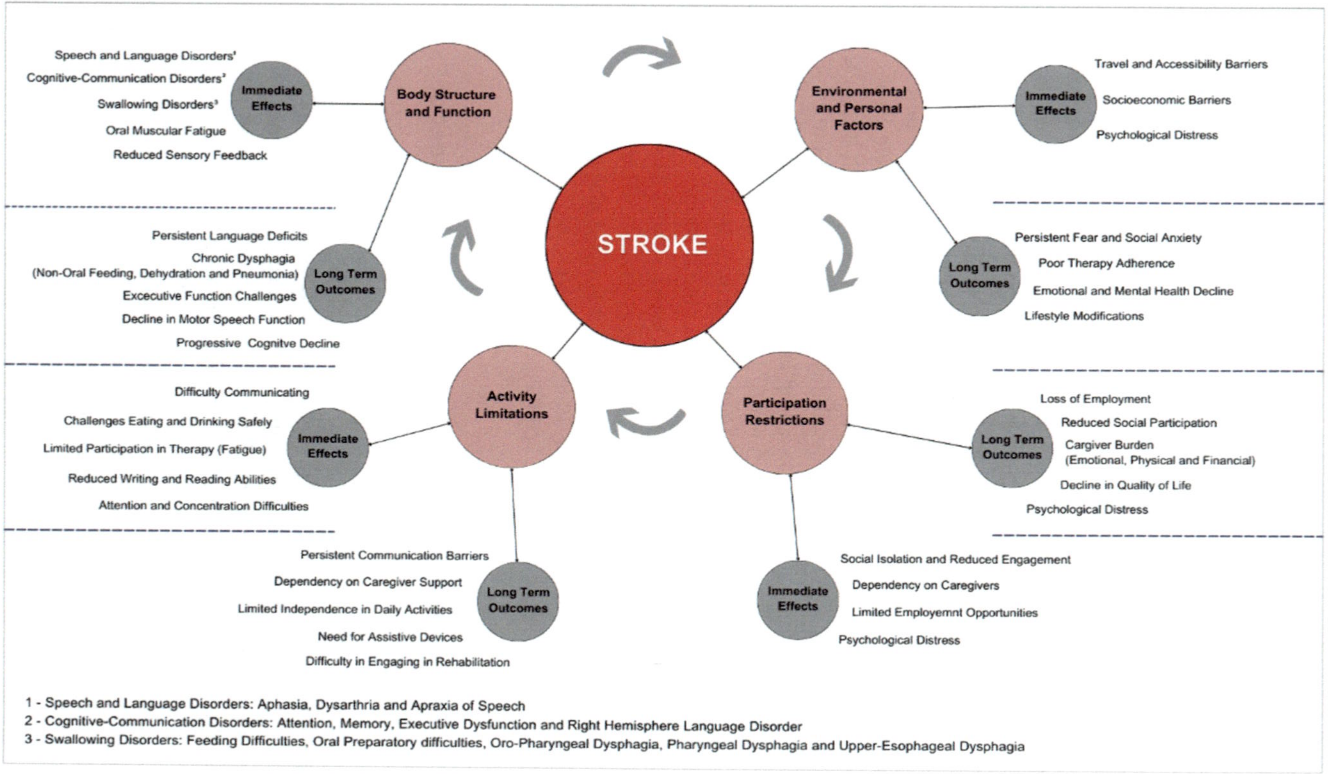

Fig. 6.1 Cyclic representation of post-stroke speech–language–swallowing impairments within the International Classification of Functioning, Disability, and Health framework

6.2.1 Neurogenic Communication Disorders

Neurogenic communication disorders following stroke include aphasia, dysarthria, apraxia of speech (AOS), and right hemisphere language disorders (RHLD). The following sections focus on rehabilitation approaches for these disorders in LMIC settings.

6.2.1.1 Aphasia

Aphasia is a acquired cognitive–linguistic disorder caused by damage to language areas in the brain [5]. Aphasia affects comprehension and expression in spoken and written modalities [26]. Considering that aphasia may co-occur with dysarthria, AOS, visual field deficits, depression, and anxiety among others, individuals with aphasia require tailored rehabilitation.

General Intervention Techniques for Aphasia Therapy

Interdisciplinary, evidence-based stroke rehabilitation led by SLPs is central to aphasia therapy [27]. Widely used behavioral therapies that enhance communication abilities and functional language use include visual action therapy, melodic intonation therapy (MIT), constraint-induced language therapy (CILT), semantic feature analysis (SFA), cueing strategies, phonological components analysis and verb network strengthening treatment, treatment of underlying forms, response elaboration training (RET), Schuell's stimulation approach, promoting aphasics' communication effectiveness, sentence production program for aphasia, voluntary control of involuntary utterances, copy and recall technique, oral reading for language in aphasia and multiple oral rereading [26], musical speech stimulation (MUSTIM), therapeutic singing (TS), and symbolic communication through music [28]. Subsequent sections adapt these techniques to the cultural and resource constraints of LMICs.

Communication Partner Training (CPT)

CPT involves caregivers, family members, friends, or volunteers to support everyday communication [26]. Partners are trained to use strategies such as using gestures, drawing, picture cards, simplified speech, clarifying questions, recapping, and reducing distractions [26, 29]. Training typically includes four steps: (a) education about aphasia and helpful strategies to improve communication, (b) raising awareness, (c) practice in real or simulated settings, and (d) feedback [26].

Group Therapy

Group therapy is useful in contexts with few SLPs or large caseloads [30]. It supports natural communication, social interaction, and psychological well-being [30], using comprehension (e.g., following directions, answering routine questions), and/or expressive tasks (e.g., turn-taking, narration, role-play). Aphasia communication groups (ACGs) extend therapy into the community through total communication methods such as drawing, gestures, writing, singing, and group conversation [31].

Adaptation of Aphasia Therapy for LMICs

Evidence-based aphasia therapy in LMICs involves using community systems, integrating cultural aspects, and employing regional resources. For instance, in MIT, SLPs can intone functional words with familiar chants or melodies that mirror speech inflections. Further, CILT can be adapted to suit the person with aphasia using images or objects from the patient's everyday environment (e.g., fruits, household tools). In SFA, the SLPs can focus on a limited set of features for the target word, with emphasis on regionally relevant items like food or daily-use objects. Visual aids, such as hand-drawn picture cards, can replace costly therapy materials as a cost-effective solution. Lastly, in RET, the clinician can incorporate familiar images or scenes, such as a regional market, a festive, or a family gathering, to promote sentence elaboration in practically relatable contexts.

Community spaces, such as schools or community halls, can be helpful to conduct group therapy activities like narration, role-playing, and functional communication games in a supportive environment. For scalability, picture cards and booklets can be developed regionally, while mobile technology can support teletherapy where smartphones are available. Training caregivers or community volunteers to learn supportive communication strategies ensures continuity of care. By adapting these techniques, aphasia therapy in LMICs becomes more practical, inclusive, and effective.

6.2.1.2 Dysarthria

Dysarthria is a motor-speech disorder caused by impaired neuromuscular control affecting the respiratory, phonatory, articulatory, resonatory, and prosodic subsystems [2]. Dysarthria leads to reduced intelligibility, naturalness, and speech efficiency [2]. Intervention techniques aim to enhance speech production through: (a) restorative approaches—improving neuromuscular function to enhance speech quality, and (b) compensatory strategies—maximizing communication despite residual deficits [32].

General Intervention Techniques for Dysarthria

Respiratory subsystem. Interventions include maximum vowel prolongation to improve breath support and control, speech-breathing training to coordinate breathing with speech tasks [2]. While inspiratory–expiratory muscle training and oral motor and respiratory exercises (OMREX) strengthen the respiratory muscles, thereby enhancing phonation and speech production [2, 28, 33], deep breathing exercises combined with posture correction further optimize respiratory efficiency [2].

Phonatory and resonatory subsystem. Lee Silverman Voice Treatment Loud [2, 32] and vocal intonation therapy (VIT) [28] improve loudness and voice quality [34]. Effortful closure techniques strengthen vocal fold adduction, enhancing vocal intensity, while sustained phonation tasks and pitch glides improve vocal endurance and flexibility [2, 32]. Hypernasality can be managed using continuous positive airway pressure therapy for velopharyngeal strengthening, along with compensatory strategies (e.g., slower rate of speech, exaggerated articulation) and palatal lift prostheses [2, 32, 35].

Articulatory subsystem and prosodic deficits. Phonetic placement, OMREX and VIT [28], intelligibility drills, and minimal pair practice improve speech clarity and precision [2]. Prosody can be enhanced through contrastive stress drills, pitch-modulation exercises, and rate-control techniques, such as pacing boards or rhythmic speech cueing (RSC) [28, 36]. For severe dysarthria, augmentative and alternative communication (AAC) systems, ranging from low-tech picture boards to high-tech speech-generating devices, provide alternative means of communication [37].

Adaptation of Dysarthria Management for LMICs

Respiratory subsystem. Household items like balloons, bottles, harmonica/flute, or whistles can be used for resistance training, while low-cost posture aids support breathing efficiency. Integrating yoga and breathing exercises into daily routines can promote consistency.

Phonatory and resonatory subsystem. Low-cost voice amplifiers can aid individuals with reduced vocal intensity. Communication health workers or caregivers can guide structured phonation tasks (sustained vowels and pitch glides) using simple auditory feedback, while visual cues (paper paddles, laryngeal mirrors) help to monitor oral and nasal resonance.

Articulatory subsystem and prosodic training. Using familiar, culturally relevant materials, phonetic placement, neurologic music therapy, and intelligibility drills can address articulation issues. Prosody can be practiced through rhythm-based exercises using regional music or songs.

AAC and Caregiver Involvement

For severe dysarthria, low tech AAC tools like picture books and alphabet or communication boards with culturally appropriate symbols help bridge communication gaps. Training caregivers in routine articulatory and intelligibility drills, supported by a native language reference manual and teletherapy, ensures accessibility and scalable intervention.

6.2.1.3 Apraxia of Speech

AOS is a motor-speech disorder marked by impaired motor planning and speech execution [22]. AOS is characterized by groping during speech, in the absence of muscle weakness, abnormal tone, or incoordination [2]. AOS affects speech clarity, naturalness, and fluency, leading to articulatory errors, rhythm disruptions, and variable tempo [2].

Improving Motor Planning and Execution in AOS

Structured, repetitive practice based on principles of motor learning are central to AOS rehabilitation [38]. Integral Stimulation Therapy, following the '*watch me, listen to me, say it with me*' approach, effectively improves motor planning and articulation [39]. Articulatory kinematic approaches, such as the eight-step continuum, sound production treatment (SPT), MIT, MUSTIM, and TS [28] initially begin with clinician–patient unison production and progress to functional word use in role-play contexts [40]. Rate- and rhythm-control techniques such as RSC [28], metronomic pacing, and pacing boards help regulate timing of speech production. For severe AOS, tactile-kinesthetic methods like prompts for restructuring oral muscular phonetic targets (PROMPT), provide hands-on support for accurate articulatory movements and sound production [41]. Intersystemic facilitation and reorganization techniques further aid speech by combining gestures or sign language with verbal attempts. Collectively, emphasizing functional communication, patient preferences, and self-monitoring support steady progress in AOS.

Adaptation of AOS Rehabilitation for LMICs

In LMICs, therapies like MIT, MUSTIM, and TS can be adapted using culturally familiar melodies. For articulatory kinematic approaches, such as SPT or the Eight-Step Continuum, SLPs can use high-frequency, environmentally relevant words. Rate and rhythm training through RSC, clapping, or metronome apps provides structured practice without advanced tools. AAC aids such as picture or alphabet boards can be made from regional materials and customized with culturally meaningful symbols for severe AOS. Intersystemic facilitation can be culturally adapted by pairing gestures with speech, while affordable tools like mirrors and minimal-pair drills (e.g., bat vs. pat) offer effective feedback and practice. Further, caregivers can use affordable tools like mirrors or flashcards and integrate practice into daily routines, such as mealtime conversations, to promote consistent engagement. Providing training materials in regional languages supports continuity of home-based exercises. Additionally, community health workers and group sessions can extend professional guidance, monitor progress, and foster motivation and social support.

6.2.1.4 Right Hemisphere Language Disorder

RHLD is an acquired condition affecting physical, behavioral, and cognitive–linguistic functions. Primary impairments include anosognosia (poor awareness of deficits), left neglect, visuospatial disorientation, constructional apraxia, amusia (difficulty recognizing musical notes), and pragmatic- or discourse-level language deficits [26, 42]. Secondary impairments such as neglect dyslexia and dysprosodia further complicates recovery in individuals with RHLD [26, 42].

Interventions for RHLD

Rehabilitation in RHLD focuses on cognitive–communication challenges, especially neglect, pragmatics, prosody, and discourse [26, 42]. Visual scanning therapy (VST) and musical neglect training (MNT) [28] are effective for left hemispatial neglect, improving visual exploration and spatial awareness through scanning and anchoring tasks [43]. Line bisection, object search, and visual or tactile anchors further enhance reading, navigation, and other functional activities. Pragmatic and psychosocial training, including musical executive function and counseling-based approaches [28, 42, 44] enhances turn-taking, topic maintenance, and interpretation of nonverbal cues, promoting more meaningful social interactions. Prosody training using VIT, RSC, pitch-intonation training, and contrastive stress drills [28, 45] restores emotional tone and expressive intonation. Inference and discourse training improve comprehension of implied meaning and figurative language through storytelling and structured discussions [26, 42]. Active tasks, like short-story analysis and structured discussions, enhance comprehension of complex information [26, 42]. Lastly, anosognosia awareness training uses feedback and self-monitoring to help patients recognize and reflect on their deficits [26].

Adaptations of RHLD Rehabilitation for LMICs

In LMICs, SLPs can adapt RHLD therapy using simple, culturally relevant tools. VST can be delivered with hand-drawn lines, colored cues, or tactile markers; pragmatic training can use culturally appropriate role-play based on daily interactions to enhance functional communication; and prosody training can incorporate regional melodies or chants for familiarity and engagement. Regional stories and folktales can aid inference and discourse work, while neglecting dyslexia can be managed with highlighted margins or simple reading charts. Spatial orientation can be reinforced by caregivers using familiar maps and landmarks. Neurologic music therapy approaches can address both motor speech and cognitive–affective deficits [28, 33]. Integrating RHLD interventions within community-based rehabilitation makes therapy more feasible and sustainable in resource-limited settings.

6.2.2 Neurogenic Dysphagia

Swallowing is a highly coordinated process involving oral, pharyngeal, and esophageal phases [46]. Dysphagia or difficulty swallowing solids and/or liquids is associated with high morbidity and mortality worldwide [47], including LMICs [48]. Hence, early swallow screening for all post-stroke patients, followed by detailed swallowing evaluation and intervention, significantly enhances swallowing safety and outcomes [49].

6.2.2.1 Dysphagia Therapy

Over the past five decades, SLP-led dysphagia therapy has evolved considerably [50]. Dysphagia therapy includes traditional and contemporary approaches. Traditional approaches include: (a) compensatory strategies, such as head positioning, sensory stimulation, and diet or texture modifications [51]; (b) muscle strengthening by targeted swallowing exercises and maneuvers [52]; and (c) neuro-muscular electrical stimulation to strengthen weak swallowing and masticatory muscles [53]. Contemporary approaches emphasize: (a) skill-based training with biofeedback to refine swallowing control [54] and (b) neuro-modulation techniques such as transcranial direct current stimulation (tDCS), repetitive transcranial magnetic stimulation (rTMS), and pharyngeal electrical stimulation [55]. Together, these approaches aim to restore safe swallowing, maintain nutrition and hydration, and reduce complications post-stroke [56, 57].

6.2.2.2 Adaptations of Dysphagia Therapy for LMICs

In LMICs, dysphagia therapy is often initiated by SLPs during hospitalization, but post-discharge care largely depends on patients and caregivers with limited resources [58]. Hence, SLPs must establish caregiver training and support systems with periodic remote monitoring. Multimedia resources (videos, illustrations, educational content) from peer-reviewed or professional platforms (e.g., MedBridge, Northern Speech Systems) help patients and caregivers understand the intricacies of swallowing, promote adherence to exercises, improve self-motivation, and catalyze the learning of exercises [59]. Regular guidance can be provided through exercise logbooks, video recordings, and telecommunication apps such as WhatsApp, Zoom, or Google Meet, allowing SLPs to monitor progress and adjust therapy remotely.

Further, low-cost innovations can substitute for expensive tools: smiley balls or elastic bands for chin-tuck exercises, three-ball respirators or balloons for expiratory muscle training, and gelatin, corn starch, or arrowroot as natural thickeners. For sensory stimulation, household items such as cold spoons, lemon juice, or tamarind can trigger swallowing reflexes, while mirrors or video playback provide visual feedback in place of biofeedback systems. By leveraging these strategies, SLPs can ensure that patients with dysphagia post-stroke receive consistent and effective care.

6.3 Practical Approaches to Addressing Speech–Language-Swallowing Rehabilitation Challenges in LMICs

Figure 6.2 summarizes the practical approaches to speech–language-swallowing rehabilitation in LMICs. The hand in the figure represents the role of SLPs within the multidisciplinary stroke team supporting and uplifting individuals with stroke

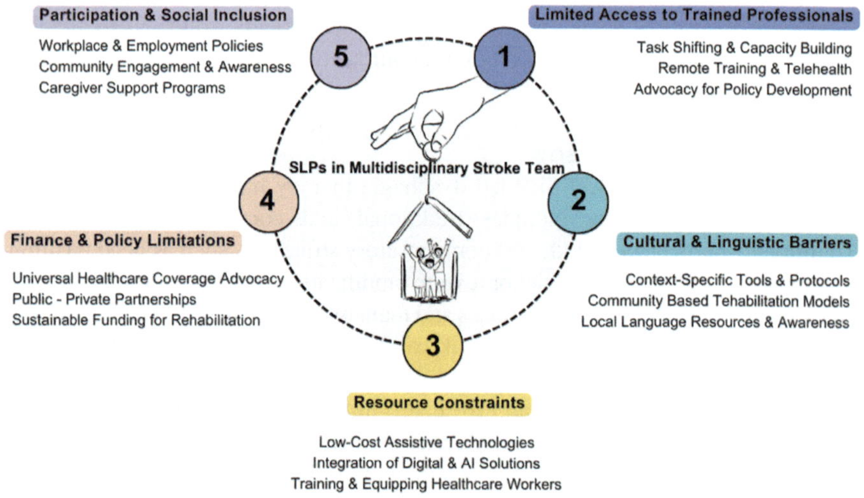

Fig. 6.2 Solutions to neurogenic communication and dysphagia-rehabilitation challenges in LMICs

through rehabilitation. Task shifting and capacity building are central strategies, where training healthcare professionals, community health workers, caregivers, and volunteers in basic rehabilitation techniques decentralize services in areas with limited SLP availability [60]. The "*train-the-trainer*" model further strengthens community-driven capacity building [61]. Additionally, telerehabilitation and low-cost mobile applications enable equitable remote guidance from SLPs [62], while advocating for mandatorily including SLP services in national stroke rehabilitation policies ensures fair access to care across LMICs [63].

6.4 Summary

Evidence-based speech–language–swallowing therapy post stroke in LMICs is crucial for improving the survivors' QoL. This chapter has highlighted the multidimensional barriers, innovative strategies, and culturally adapted interventions that can bridge gaps in rehabilitation services across LMICs. Strategies such as task-shifting, capacity building, and the use of low-cost, technology-driven approaches are key to overcoming resource constraints and ensuring sustainable rehabilitation. Moreover, SLPs can optimize outcomes even in low-resource settings by integrating principles of neural plasticity, motor learning, and exercise physiology into rehabilitation frameworks. Community engagement, caregiver training, and interdisciplinary teamwork further enhance the rehabilitation effectiveness, ensuring that individuals with neurogenic communication disorders and dysphagia receive timely, comprehensive, and context-specific care. Future initiatives should focus on developing regional validated assessment tools, strengthening professional training programs, and advocating for policies that integrate speech–language pathology services into primary healthcare systems.

What Does Evidence Say [64]

- Aphasia: Therapy should start within the first 4 weeks after stroke to optimize recovery. In the acute phase (up to 6 weeks), sessions of 30–45 min, 2–3 times per week are appropriate, with limited added benefit from extra sessions. In the chronic phase (>6 months), intensive therapy (≥10 h/week of therapist-led sessions for 3 weeks, plus ≥5 h/week of self-practice) can improve outcomes. Therapy should address functional communication, comprehension (reading and auditory), expressive language, and writing.
- Dysphagia: Swallowing should be screened using validated tools within 4 h of hospital admission and before any oral intake. Ongoing management should include direct therapy (skill and strength training), indirect strategies (dietary modifications and safe swallowing advice), and maintenance of good oral hygiene.
- Apraxia of speech (AOS): Treatment should be individualized and may include articulatory–kinematic and rate/rhythm approaches. Techniques can involve modeling, visual cues, motor learning principles, PROMPT, and

(continued)

computer-based multimodal practice. For functional communication, AAC methods such as gestures or speech-generating devices are recommended.

- Dysarthria: Management should be tailored to the individual and focus on improving connected speech through strategies like slowing speech rate, enhancing articulation, and increasing vocal loudness.
- Communication partner training (CPT): Training should be provided to healthcare staff and volunteers, and can also be extended to caregivers of individuals with post-stroke aphasia.
- Brain stimulation: Techniques such as tDCS or rTMS are not recommended for routine clinical use in chronic aphasia and should be limited to research settings.

References

1. Behera DK, Rahut DB, Mishra S. Analyzing stroke burden and risk factors in India using data from the global burden of disease study. Sci Rep. 2024;14(1):1–18.
2. Duffy JR. Motor speech disorders: substrates, differential diagnosis, and management. 4th ed. Amsterdam: Elsevier; 2019.
3. Wray F, Clarke D, Forster A. How do stroke survivors with communication difficulties manage life after stroke in the first year? A qualitative study. Int J Lang Commun Disord. 2019;54(5):814–27.
4. Sheppard SM, Sebastian R. Diagnosing and managing post-stroke aphasia. Expert Rev Neurother. 2021;21(2):221–34.
5. Mesulam MM. Aphasia, sudden and progressive. In: Whitaker HA, editor. Concise encyclopedia of brain and language. Oxford: Elsevier; 2010. p. 49–53.
6. Cichero JA, Murdoch BE. Dysphagia: foundation, theory and practice. Hoboken: John Wiley & Sons; 2006.
7. Surya N, Kumar S, Someshwar H, Srivastava A, Gandhi D, Palande D, et al. Advancing neurorehabilitation in India and key steps to reach WHO'S rehab 2030 goals. 2018.
8. Tiwari S, Joshi A, Rai N, Satpathy P, Tiwari S, Joshi A, et al. Impact of stroke on quality of life of stroke survivors and their caregivers: a qualitative study from India. J Neurosci Rural Pract. 2021;12(4):680–8.
9. Langton-Frost N, Brodsky MB. Speech-language pathology approaches to neurorehabilitation in acute care during COVID-19: capitalizing on neuroplasticity. PM&R. 2022;14(2):217–26.
10. Albin M, Wiseman-Hakes C. Cognitive-communication disorders and neurodisability in the criminal justice system: emerging roles, responsibilities, and opportunities for speech-language pathologists. Perspect ASHA Spec Interes Groups. 2023;8(6):1304–20.
11. Karanth P. Four decades of speech-language pathology in India: changing perspectives and challenges of the future. Folia Phoniatr Logop. 2002;54(2):69–71.
12. Mandke K, Bellur R. Genesis of the speech and hearing profession in India. Manipal: Manipal University Press; 2024.
13. Abou-Elsaad T. Training and services for the communication and swallowing disorders in Egypt: a short report. Arab J Appl Linguist. 2022;7(1):99–105.
14. Cheng WT, Olea TCM, Marzan JCB. Speech-language pathology in The Philippines: reflections on the past and present, perspectives for the future. Folia Phoniatr Logop. 2002;54(2):79–82.
15. Charalambous M, Pierce JE, Pastou G, Kola E, Savitz SI. The role of speech and language therapists (SLTs) in international stroke teams: a systematic review. Int J Lang Commun Disord. 2025;60(4):e70062.

16. Kayola G, Mataa MM, Asukile M, Chishimba L, Chomba M, Mortel D, et al. Stroke rehabilitation in low- and middle-income countries: challenges and opportunities. Am J Phys Med Rehabil. 2023;102(2):S24–32.
17. Lokheshwar S, Rajasudhakar R. Medical practitioner's awareness on speech-language pathologists and their services in India: a pilot study. J Interprofessional Educ Pract. 2021;23:100412.
18. Gardner D, Mitchell V, Frakking T, Weir KA, Canning A, Wenke RJ. Validated patient reported outcome measures in speech-language pathology: a scoping review of adult practice. Int J Speech Lang Pathol. 2024;27(5):672–701.
19. Rangarathnam B, Desai R. A preliminary survey of dysphagia practice patterns among speech-language pathologists in India. J Indian Speech Lang Hear Assoc. 2020;34(2):259.
20. Muttiah NA, McNaughton D, Drager KDR. Providing instructional support for AAC service delivery in low- and middle-income (LAMI) countries. Int J Speech Lang Pathol. 2016;18(4):341–53.
21. Kleim JA, Jones TA. Principles of experience-dependent neural plasticity: implications for rehabilitation after brain damage. J Speech Lang Hear Res. 2008;51(1):S225–39.
22. Maas E, Robin DA, Wulf G, Austermann Hula SN, Freedman SE, Wulf G, et al. Principles of motor learning in treatment of motor speech disorders. Am J Speech-Language Pathol. 2008;17(3):277–98.
23. Stathopoulos E, Ducchan JF. History and principles of exercise-based therapy: how they inform our current treatment. Semin Speech Lang. 2006;27(4):227–35.
24. Ward NS, Cohen LG. Mechanisms underlying recovery of motor function after stroke. Arch Neurol. 2004;61(12):1844–8.
25. Schmidt RA, Lee TD. Motor learning and performance: from principles to application. 5th ed. Champaign: Human Kinetics; 2014.
26. Albys DG. Aphasia and related cognitive-communicative disorders. Boston: Pearson; 2014.
27. Burton B, Isaacs M, Brogan E, Shrubsole K, Kilkenny MF, Power E, et al. An updated systematic review of stroke clinical practice guidelines to inform aphasia management. Int J Stroke. 2023;18(9):1029–39.
28. Thaut MH, Hoemberg V. Handbook of neurologic music therapy. Oxford: Oxford University Press; 2014.
29. Matos MAC, Abreu E, Brandão L, Santos JMLG, Lima RR, Machado TH, et al. Communication partner training for persons with aphasia: a systematic review of Portuguese language publications. Aphasiology. 2024;39:539–78.
30. Layfield CA, Ballard KJ, Robin DA. A scholarly forum for guiding evidence-based practices in speech-language pathology evaluating group therapy for aphasia: what is the evidence? EBP Briefs. 2013;7(5):1–17.
31. Charalambous M, Kambanaros M. The importance of aphasia communication groups. In: Jianu DC, Mureşanu DF, editors. Aphasia compendium. London: IntechOpen; 2021.
32. Freed DB. Motor speech disorders: diagnosis and treatment. 3rd ed. San Diego: Plural Publishing; 2020.
33. Thaut MH, Hegde S, Akkunje PS. Neurologic music therapy: applied music neuroscience for neurorehabilitation and global brain health. In: Surya N, editor. IFNR textbook of neurorehabilitation. 1st ed. Evangel Publications: New Delhi; 2024. p. 484–8.
34. Yuan F, Guo X, Wei X, Xie F, Zheng J, Huang Y, et al. Lee Silverman voice treatment for dysarthria in patients with Parkinson's disease: a systematic review and meta-analysis. Eur J Neurol. 2020;27(10):1957–70.
35. Cahill L, Turner AB, Stabler PA, Addis PE, Theodoros DG, Murdoch BE. An evaluation of continuous positive airway pressure (CPAP) therapy in the treatment of hypernasality following traumatic brain injury a report of 3 cases. J Head Trauma Rehabil. 2024;19(3):241–53.
36. Tamplin J, Grocke D. A music therapy treatment protocol for acquired dysarthria rehabilitation. Music Ther Perspect. 2008;26(1):23–9.
37. Elsahar Y, Hu S, Bouazza-Marouf K, Kerr D, Mansor A. Augmentative and alternative communication (AAC) advances: a review of configurations for individuals with a speech disability. Sensors. 2019;19(8):1911.

38. Munasinghe TU, Ariyasena ADK, Siriwardhana DD. Speech therapy interventions for acquired apraxia of speech: an updated systematic review. Am J Speech-Language Pathol. 2023;32(3):1336–59.
39. Gildersleeve-Neumann C. Treatment for childhood apraxia of speech: a description of integral stimulation and motor learning. ASHA Lead. 2007;12(15):10–30.
40. Bailey DJ, Eatchel K, Wambaugh J. Sound production treatment: synthesis and quantification of outcomes. Am J Speech-Language Pathol. 2015;24(4):S798–814.
41. Bose A, Square PA. PROMPT treatment method and apraxia of speech. Perspect Neurophysiol Neurogenic Speech Lang Disord. 2001;11(4):5–8.
42. Brookshire RH, McNeil MR. Introduction to neurogenic communication disorders. 8th ed. Amsterdam: Elsevier; 2014.
43. Van Wyk A, Eksteen CA, Rheeder P. The effect of visual scanning exercises integrated into physiotherapy in patients with unilateral spatial neglect poststroke: a matched-pair randomized control trial. Neurorehabil Neural Repair. 2014;28(9):856–73.
44. Blake ML. Communication deficits associated with right hemisphere brain damage. In: Damico JS, Muller N, Ball MJ, editors. The handbook of language and speech disorders. Hoboken: John Wiley & Sons; 2021. p. 571–89.
45. Benedetti V, Weill-Chounlamountry A, Pradat-Diehl P, Villain M. Assessment tools and rehabilitation treatments for aprosodia following acquired brain injury: a scoping review. Int J Lang Commun Disord. 2022;57(3):474–96.
46. Matsuo K, Palmer JB. Anatomy and physiology of feeding and swallowing: normal and abnormal. Phys Med Rehabil Clin N Am. 2008;19(4):691–707.
47. Banda KJ, Chu H, Kang XL, Liu D, Pien LC, Jen HJ, et al. Prevalence of dysphagia and risk of pneumonia and mortality in acute stroke patients: a meta-analysis. BMC Geriatr. 2022;22(1):1–10.
48. Krishnamurthy R, Balasubramanium RK, Premkumar PK. Systematic review and meta-analysis of dysphagia and associated pneumonia in patients with stroke from India: a call to arms. Am J Speech-Language Pathol. 2022;31(1):502–14.
49. Sherman V, Greco E, Martino R. The benefit of dysphagia screening in adult patients with stroke: a meta-analysis. J Am Heart Assoc. 2021;10(12):18753.
50. Huckabee ML, Mills M, Flynn R, Doeltgen S. The evolution of swallowing rehabilitation and emergence of biofeedback modalities. Curr Otorhinolaryngol Rep. 2023;11(2):144–53.
51. Lazarus CL. History of the use and impact of compensatory strategies in management of swallowing disorders. Dysphagia. 2017;32(1):3–10.
52. Choy J, Pourkazemi F, Anderson C, Bogaardt H. Dosages of swallowing exercises in stroke rehabilitation: a systematic review. Eur Arch Oto-Rhino-Laryngol. 2023;280(3):1017–45.
53. Wang Z, Xiao Z, Shen Q, Zhao N, Zhang W. Neuromuscular electrical stimulation for post-stroke dysphagia treatment: a systemic evaluation and meta-analysis of randomized controlled trials. Dysphagia. 2024;39(3):424–32.
54. Benfield JK, Woods K, Woodhouse L, Roadevin C, James M, Bath PM, England TJ, Radford K, Steele C. Swallow Strength and Skill Training with biofeedback In cute Post stroke dysphagia (ssSIP): a protocol for a multi-centre feasibility trial. Pilot Feasibility Stud. 2026.
55. Cheng I, Sasegbon A, Hamdy S. Evaluating the therapeutic application of neuromodulation in the human swallowing system. Dysphagia. 2022;38(4):1005–24.
56. Johnson DN, Herring HJ, Daniels SK. Dysphagia management in stroke rehabilitation. Curr Phys Med Rehabil Rep. 2014;2(4):207–18.
57. Zuckerman M, Wang S, Kaneoka A, Coster WJ, Leonard R, Langmore SE, et al. Conceptualizing adult dysphagia in the United States within the international classification of functioning, disability and health (ICF). Arch Phys Med Rehabil. 2024;105(5):1008–18.
58. Jayes M, Madima V, Marshall J, Pillay M. Dysphagia management in community/home settings: a scoping review investigating practices in Africa. Int J Speech Lang Pathol. 2024;26:861–72.

59. Govender R, Taylor SA, Smith CH, Gardner B. Helping patients with head and neck cancer understand dysphagia: exploring the use of video-animation. Am J speech-language Pathol. 2019;28(2):697–705.
60. Leong SL, Teoh SL, Fun WH, Lee SWH. Task shifting in primary care to tackle healthcare worker shortages: an umbrella review. Eur J Gen Pract. 2021;27(1):198–210.
61. Habibi J, Bosch J, Bidulka P, Belson S, DePaul V, Gandhi D, et al. Strategies for specialty training of healthcare professionals in low-resource settings: a systematic review on evidence from stroke care. BMC Med Educ. 2023;23(1):1–15.
62. Srivastava A, Swaminathan A, Chockalingam M, Srinivasan MK, Surya N, Ray P, et al. Tele-neurorehabilitation during the COVID-19 pandemic: implications for practice in low- and middle-income countries. Front Neurol. 2021;12:667925.
63. Kamalakannan S, Srivastava A, Chockalingam M, Gandhi D, Dhamija RK, Solomon J, et al. Health systems for rehabilitation—critical appraisal of the national programme for stroke in India. J Indian Fed Neurorehabil. 2024;1(1):26.
64. Stroke Foundation. Living clinical guidelines for stroke management [internet]. 2025. Available from: https://informme.org.au/guidelines/living-clinical-guidelines-for-stroke-management.

Contextualized Activities of Daily Living in Stroke Rehabilitation

7

Premraj Issac Chandran,
Sultan Salem Hammam Alabdulla, Thajus Asirvatham,
and Sureshkumar Kamalakannan

Abstract

This chapter explores the background and significance of disability following stroke, outlining the purpose and scope of the discussion to highlight the challenges faced by stroke survivors and the need for effective rehabilitation strategies. The chapter particularly focuses on activities of daily living (ADLs), distinguishing between basic and instrumental ADLs, and discusses approaches to retraining these skills and their effects. Instrumental ADLs are given special attention due to their importance in promoting independence. It also highlights the importance of contextualized rehabilitation tailored to individual needs, identifies gaps in current practice, and supports these insights with relevant literature, underscoring the vital role of personalized rehabilitation in enhancing the quality of life for stroke survivors.

P. I. Chandran · S. S. H. Alabdulla · T. Asirvatham
Department of Occupational Therapy, Qatar Rehabilitation Institute, Hamad Medical Corporation, Doha, Qatar

S. Kamalakannan (✉)
Faculty of Health and Life Sciences, Department of Social Work, Education, and Community Wellbeing, Northumbria University, Newcastle upon Tyne, UK
e-mail: sureshkumar.kamalakannan@northumbria.ac.uk

D. B. C. Gandhi, J. D. Pandian (eds.), *Stroke Rehabilitation in Low-Resource Settings*, https://doi.org/10.1007/978-981-92-0403-8_7

7.1 Introduction

7.1.1 Background and Significance

Stroke continues to be the leading cause of disability worldwide, with an estimated 12.2 million new cases annually [1]. The growing burden of stroke and its associated care presents a major public health challenge, particularly in low- and middle-income countries (LMICs), where access to stroke care and rehabilitation remains limited [2]. With the increase in stroke cases, there is an imperative need for treatment and intervention to treat the underlying pathology and disease [3]. While medical management—such as anticoagulants, blood thinners, or procedures like thrombectomy—addresses the underlying pathology, rehabilitation is essential for recovery and reintegration into premorbid lifestyles [4]. Stroke rehabilitation is a multidisciplinary approach that enables survivors with physical, cognitive, psychological, social, or vocational impairments to maximize functioning within environmental constraints [5]. It is therefore an overarching goal of rehabilitation to remediate and compensate for post-stroke limitations and disability [6].

7.1.2 Purpose of the Chapter

The primary goal of rehabilitation is to regain lost function and independence, and improve the overall quality of life, by reducing the limitations of activities of daily living (ADLs) to a minimum [7]. A structured stroke programme should consist of the timing of rehabilitation provided, and a qualified multidisciplinary team including a physiatrist, rehabilitation nurse, physiotherapist, occupational therapist, speech-language pathologist, psychologist, social worker, and many more [8]. The duration, functionality, and willingness of the stroke survivor determine the overall outcomes of the stroke programme and the recovery path of the affected individual [9]. An organized and well-structured rehabilitation programme and adequate intensity of rehabilitation services are associated with better and improved functional outcomes in stroke survivors than ordinary stroke units or other programmes such as mobile stroke units or general units [10]. Thus, stroke rehabilitation is most important to retrain and facilitate functional outcomes that primarily target ADL.

7.2 Understanding Disability Following Stroke

Disability is defined as an umbrella term that incorporates impairments, activity limitations, and participation restrictions, according to the International Classification of Functioning Disability and Health (ICF) [11]. The ICF describes disability as a complex process reflecting the interaction between a stroke survivor (in our case) and the environment in which he/she experiences a stroke. For example, the stroke survivor may have hand function problems (impairment) that could reduce his/her opportunity to participate in his/her self-care activities like dressing,

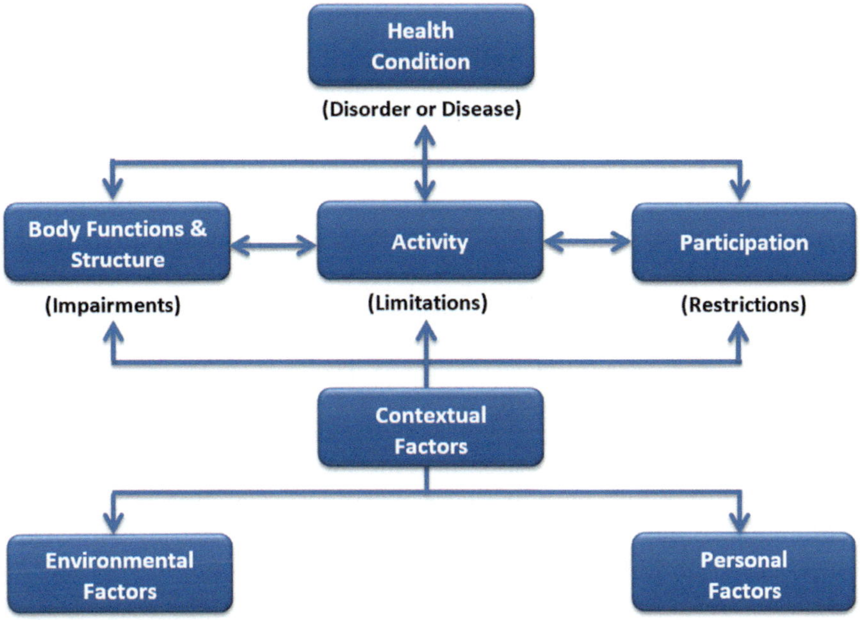

Fig. 7.1 The international classification of functioning, disability and health, World Health Organization (WHO)

bathing, grooming, feeding, etc. (disability). However, if the individual is provided with assistive technology, he or she will be able to independently participate in most of their basic activities of daily living. Figure 7.1 illustrates the disability and functioning framework of any health condition through the ICF.

Stroke-related disability is not just related to visible or invisible impairments. It is the subsequent interactions of the person experiencing an impairment and the environment in which the experience occurs [12]. The severity of disability depends on the degree of impairment (physical, mental, cognitive) as well as the personal and contextual environment of the affected individual [12].

7.2.1 Rehabilitation Needs and Everyday Life of the Stroke Survivors

Stroke survivors experience different kinds of needs due to the disabling nature of the brain injury that obstructs their independent engagement in everyday living [13]. Due to post-stroke disabilities, stroke survivors may require support to engage in activities of daily living, such as feeding, bathing, dressing, toileting, and mobility [14]. Coping with the sudden onset of a diverse unforeseen disability is another important issue that impacts the everyday life of a stroke survivor, in addition to their physical needs [15]. The mental health post-stroke gets severely affected in most individuals and presents as depression and anxiety, requiring psychological support [16].

Coping with the sudden changes following a stroke is an important rehabilitation need for stroke survivors [17]. Similarly, stroke survivors may also require assistance with returning to their productive employment and participating in social situations like searching for a new job or managing family responsibilities like childcare and financial management [18]. These needs arise when the stroke survivor gets medically stabilized and returns home [19]. Not all stroke survivors have the same needs. It differs based on the phase of recovery as well as the severity of their disability [20]. If the ICF framework is considered in this situation, the rehabilitation needs are also determined by the environment in which the stroke survivors experience disability, which also becomes critical (e.g., accessibility to rehabilitation services, socio-economic status, family support, etc.).

Therefore, stroke recovery and rehabilitation are an important need for any stroke survivor [21]. Effective rehabilitation is a systematic process involving a thorough, comprehensive assessment of the needs of the individual, setting specific goals that are relevant to the felt needs to enable the stroke survivor to achieve independence in their activities of daily living and enhance their continuum of recovery and care effectively [22]. This cannot be achieved by just a single rehabilitation expert but a comprehensive, multidisciplinary rehabilitation team such as physiotherapists, occupational therapists, speech therapists, rehabilitation physicians, rehabilitation nurses, clinical psychologists, social workers, and assistants with expertise in their respective professional disciplines [23]. There are several ways to provide good quality rehabilitation and care, but the best way is for these members of the rehabilitation team to work together, complementing each other in achieving therapeutic goals set for the stroke survivor. The evidence-based guidelines for the provision of various rehabilitation interventions for stroke survivors and the recommended clinical guidelines to promote regaining independence in activities of daily living are provided in Table 7.1.

Table 7.1 Stroke rehabilitation guidelines and recommendations for recovery of activities of daily living [21]

Interventions	Goal	Recommendation category	Grade of recommendation
Beneficial or likely to be beneficial			
Multidisciplinary stroke-unit care	To improve independence	Recommended	A
Early supported discharge services	To improve independence	Recommended	A
Therapy-based rehabilitation services at home (within 1 year of stroke)	To improve ADL	Recommended	A, B
Outpatient rehabilitation services (day-hospital, community team)	To improve ADL	Selected use	A, B
Rehabilitation services in long-term care settings	To improve ADL	Selected use	B

(continued)

Table 7.1 (continued)

Interventions	Goal	Recommendation category	Grade of recommendation
Occupational therapy services	To improve ADL	Recommended	A, B
Occupational therapy services at home	To improve ADL and extended ADL	Recommended	A
Uncertain benefits			
Integrated-care pathways	To improve independence	Not recommended or Selected use	B
Services with stroke liaison workers and participation	To improve independence	Not mentioned	
Information provision	To improve knowledge and independence	Recommended	A
Therapy-based rehabilitation services at home (after 1 year)	To improve ADL	Selected use	B, C

Guideline recommendation categories: recommended = recommended use for a substantial proportion of stroke patients; selected use = might be considered in selected patients or circumstances, not mentioned = no specific recommendation made; not recommended = not recommended for routine use (outside the context of a clinical trial)
Guideline grade of recommendation categories: (A) based on robust information from randomised trials that is applicable to the target population; (B) based on less robust information (from experimental studies); and (C) consensus or expert opinion. *ADL* activities of daily living

7.3 Activities of Daily Living and Stroke

7.3.1 Basic and Instrumental ADL

Activities of daily living are diverse, as are the core concepts of any stroke rehabilitation programme [24]. Occupational therapists are the specialized experts in enabling independence in ADL among individuals with disabilities in general [24]. ADL are daily household-based activities or occupations that people carry out to maintain health and well-being. There are two different types of ADL. The first kind of ADL is called basic ADL (BADL), and the second kind is instrumental ADL (IADL).

The basic ADLs include (a) mobility: the extent of an individual's ability to move from one position to another and walk independently; (b) eating: the ability of a person to feed oneself; (c) dressing: the ability to select appropriate clothes and to put the clothes on; (d) personal hygiene: the ability to bathe and groom oneself and maintain dental hygiene, nail, and hair care; (e) continence: the ability to control bladder and bowel function; and (f) toileting: the ability to get to and from the toilet, use it appropriately, and clean oneself [25].

Instrumental ADLs include (a) transportation and shopping: ability to procure groceries, attend events, and manage transportation, either via driving or by organizing other means of transport; (b) managing finances: this includes the ability to pay bills and manage financial assets; and (c) shopping and meal preparation, that is everything required to get a meal on the table. It also covers shopping for clothing and other items required for daily life; (d) housecleaning and home maintenance: cleaning kitchens after eating, maintaining living areas reasonably clean and tidy, and keeping up with home maintenance; (e) managing communication with others: the ability to manage telephone and mail; (f) child care: supporting children in the family; and (g) managing medications: ability to obtain medications and take them as directed [26].

The dependence on BADL is considered a short- and long-term consequence of the debilitating stroke condition and requires specialized services such as stroke rehabilitation. It is seen that very early dependency in BADL as early as 2 days can predict and explain dependency at the end of 3–6 months post-stroke. Therefore, early assessments can provide a better understanding of rehabilitation needs following stroke [26].

7.3.2 Retraining ADL and Its Effects

In addition to assessments, the rehabilitation intervention phase needs to be adjusted to appropriately retrain stroke survivors with the needed ADL skills to enable them to be as functional as possible in their immediate world. There always exists a conflicting area of concern about the setting and where a stroke survivor should receive therapy in order to gain optimum benefit. It could be the contextually appropriate environment or that of the traditional therapy or rehabilitation setting. It is often seen that stroke survivors who were trained in simulated or contextually appropriate environments show better recovery patterns and functional improvement. However, there is a paucity of evidence in this area [27]. Simulated functional tasks or programmes may provide added benefits to overall functional gain and improvements. One study showed that such simulated functional task training improved cognition, ADL function, and functional balance among elderly persons with cognitive decline [28]. Therefore, it is rather interesting to train people in simulated environments so that they get a feel for what real-world scenarios may look like. This could also happen during acute and chronic rehabilitation in the hospital wards. This kind of simulated training in ADL could provide the stroke survivors with a sense of confidence to handle real-life scenarios as independently as possible in areas of both BADL and IADL.

Previous literature has stated a strong correlation between ADL and the quality of life (QoL) of stroke patients [29]. With stroke patients having functional limitations that may limit their ability to participate in ADL, it is seen that physical limitations and impairments are major determinants of QoL. The focus on QoL has always been highlighted as the goal of rehabilitation of stroke patients, which is to improve their QoL [30]. However, there are a few factors related to ADL deterioration in stroke patients that need to be addressed. Age, vascular risk factors, balance, and post-stroke depression are a few of those factors that are associated with ADL deterioration [31]. It is also noted that those stroke patients with recurrent stroke, particularly haemorrhagic stroke, who are dependent on financial resources have reduced nutritional status, and those living with family have poorer stroke-related outcomes [32]. Hence, such risk factors and determinants need to be keenly looked upon to further engage stroke patients to their fullest capacity in rehabilitation.

7.3.3 Instrumental ADL: An Overview

Another aspect of ADL is instrumental activities of daily living (IADL). IADLs are essential for independent living and participation in the community as they are considered an important indicator of functional independence and competence, which influence overall participation and well-being in stroke patients [33]. Limitations in IADL, such as recreation, housekeeping, hobbies, and social interaction, may affect the social participation of stroke patients in the community [34]. Previous literature has also explored predictors of IADL performance among stroke patients. It is noted that basic ADL were one of the strongest predictors of IADL performance [35]. The interactions between basic ADL and IADL among stroke patients are therefore essential in goal setting and in the rehabilitation process.

The levels of severity among stroke patients and ADL performance are related. The prognosis of rehabilitation and the levels of independence depend on stroke severity levels [36]. To those with severe functional limitations or impairments present with higher levels of dependence, retraining in BADL remains the main rehabilitation goal. With rehabilitation facilitating function, the ability to perform ADL may also improve owing to overall functional independence and recovery [37]. In contrast, those with mild to moderate severity of stroke present with better levels of independence and are able to perform BADL with varying levels of assistance. Mild and moderately disabled stroke survivors engage better in occupations and activities within the home and will require rehabilitation to function independently in the community; hence, rehabilitation support towards IADL is required for these individuals aimed at participating better in social activities [38].

Another aspect of looking into ADL, particularly IADL, is leisure participation. Engaging in leisure activities is directly related to the extent of stroke and its aftereffects. Previous studies have shown that there has been a reduction in leisure engagement following an episode of stroke, which can be related to the sociodemographic characteristics of the stroke survivor and their disability severity [39]. An important aspect of IADL rehabilitation, especially work, leisure, and home management activities among stroke survivors, is post-stroke depression and its impact on fatigue. Both these debilitating components may play a major role in participating in IADL, including leisure and work. This is especially the case among community-dwelling stroke survivors [40].

In addition to rehabilitation and recovery with professionals, the use of technology in adapting ADL among stroke survivors has been a critical area to bridge the gaps in access and to promote functional recovery. It is noted that usually by the end of the first year post-stroke, half of stroke survivors still need ADL assistance [41]. The introduction and use of technical innovations and assistive technology (AT) provide a platform and opportunity to compensate for the loss of function and facilitate ADL at an early stage, and simultaneously retrain or remediate the impairments resulting from stroke in the same individual [42]. AT is defined as any electrical or mechanical device designed to facilitate movement and recovery by offering direct assistance [43]. With higher precision and assisting in measuring ADL among stroke survivors, wearable technologies also play an important role in the field of stroke rehabilitation. Through machine-learning algorithms and sensors, wearable technologies are able to measure the extent and amount of ADL functions among stroke survivors [44]. Technologies to facilitate participation of stroke survivors in their home environment have a great potential to facilitate recovery (BADL). Technologies such as virtual reality, digital assistants, and apps can facilitate active participation of stroke survivors in the community (IADL) [45]. With rising advancements in technology, it is imperative that new, patient-centered, customized technologies can bring about a positive change in function and recovery, enhancing the abilities of the stroke survivor to engage effectively in ADL.

In summary, following a stroke, areas of BADL and IADL are compromised to varying extents depending on the severity of brain damage, disability, and the context in which one experiences it. The importance of rehabilitating those individuals based on their felt need and deciding on a therapeutic goal and plan to optimize all possible opportunities to enable function and independence in ADL. A comprehensive, multidisciplinary, inclusive strategy targeting ADLs can certainly enhance the recovery process and also reduce the complexities of care and support from the family. This approach to rehabilitation and recovery in stroke can ultimately have a very positive impact and enhance successful community re-integration of the stroke survivor, particularly in low-resource settings.

7.4 Contextualized Rehabilitation for Activities of Daily Living

7.4.1 Overview

According to the ICF framework by the WHO, it is evident that the environment plays a crucial role in the rehabilitation of a person with any kind of disability. Rehabilitation depends on not just the impairments but also activity limitations and participation restrictions. More importantly, the environment in which one experiences disability becomes a deciding factor for determining the rehabilitation goals and enabling functional independence in everyday living among those affected [11]. Therefore, contextualizing rehabilitation following a stroke for those affected by stroke-related disability is of very high importance to achieve global targets such as

Fig. 7.2 How individuals eat in different contexts (the United Kingdom on the left; India on the right)

the universal health coverage (UHC), sustainable development goals (SDGs) and rehabilitation 2030 [46]. It is also equally important to enable a stroke survivor become a productive member of the family and society in a context where he or she experiences stroke.

Contextualized rehabilitation for ADL becomes critical because context is not just defined by the geography but also the culture of the individuals experiencing stroke, and also the culture of the health and rehabilitation or social care systems that are available and accessible to these individuals affected by stroke [47]. For example, the way people eat in British and India is very different. British people will usually use a knife and fork for eating, unlike the Indians who use their fingers to eat. Please find Fig. 7.2.

Similarly, looking at the health and social care contexts in these same geographies. The United Kingdom has well-organized stroke care pathways at all levels of care, including in the communities. However, in a context like India, rehabilitation is still not yet integrated into the health systems and policies, particularly for stroke care [48]. This implies the importance of contextualizing stroke rehabilitation, particularly ADL, for the stroke survivors at all levels of stroke care and recovery.

7.4.2 Identified Gaps in Contextualization

There are many obvious gaps in contextualizing ADL within stroke care in any context. The primary gap is the understanding of regaining the function to independently engage in ADL in the communities of the affected individuals. Even in the context where stroke rehabilitation pathways are well-organized, enabling a stroke survivor and their families to regain function and independently perform their ADLs in their own communities is complex and challenging. Let alone a context where there aren't any services for stroke rehabilitation [49].

Stroke disability requires continuum of care to enable stroke survivors to independently perform their BADL as well as IADL in their own context. Promoting access to stroke care in the actual environment of the affected individual is crucial to optimizing stroke rehabilitation provided at hospitals and rehabilitation centres, and to make use of all the potential gains made at these institutionalized contexts. Therefore, stroke rehabilitation needs to be contextualized [50].

In summary, contextualized ADL as a part of a stroke rehabilitation programme is of utmost importance for both stroke survivors and their families. As stroke impacts not just the survivor but also the immediate family members, it poses a significant burden on the active engagement of those affected in their ADLs [50, 51]. Stroke creates a double burden when it comes to ADLs. This has a very clear implication for stroke care providers and policymakers in any context. Patient-centred stroke care must be comprehensive and inclusive. The felt needs of the patients as well as their context must be carefully considered in goal planning and implementation of interventions to achieve it [52]. The evidence for such interventions targeting the real-life context of the affected individuals will enhance rehabilitation and recovery in stroke. Therefore, efforts must align with generating evidence for those contextualized ADLs and stroke care in any context. Development of stroke care policies must include rehabilitation for ADLs within the content, and it must be implemented comprehensively with an inclusive workforce that can potentially enable a stroke survivor to independently engage in their ADLs in the communities [53]. The implementation must be adequately resourced with a multidisciplinary stroke rehabilitation workforce, including occupational therapists who are the pillars within stroke care and rehabilitation to promote engagement and participation in everyday activities. Stroke survivors have always felt the need to reengage in their everyday living as a primary goal in rehabilitation, and it is very important that these needs are integrated into stroke care and rehabilitation [54]. To make this successful, contextualization is one of the potential strategies. Contextualized stroke rehabilitation, particularly targeting the activities of daily living of stroke survivors, can certainly meet their needs and reduce the unmet need for rehabilitation following stroke worldwide.

7.4.3 Supporting Literature

What Does Evidence Say

- Community-dwelling stroke survivors who have difficulty performing daily activities should be assessed by a trained clinician. If impairments in personal or instrumental activities of daily living are identified, they should receive targeted therapy (e.g., task-specific training and use of assistive devices).
- Virtual reality-based interventions may be used alongside standard therapy to improve performance in activities of daily living.
- Routine occupational therapy is not recommended solely to improve activities of daily living in older stroke survivors residing in nursing homes.
- Acupuncture and the use of amphetamines are not recommended for improving activities of daily living.
- Selective serotonin reuptake inhibitors (SSRIs) should not be used for the purpose of reducing disability.

(continued)

- Brain stimulation techniques (e.g., transcranial direct current stimulation or repetitive transcranial magnetic stimulation) should not be used in routine clinical practice to improve daily functioning and should be limited to research settings.
- Targeted occupational therapy programmes, including leisure-focused interventions, may be used to enhance participation in leisure activities.
- All stroke survivors should be asked about their pre-stroke employment (paid or unpaid) and their desire to return to work. Those interested should undergo assessment of their abilities relative to job demands and be offered support such as workplace interventions, worksite visits, or referral to supported employment services.

Copyright Disclaimer

References

1. GBD 2019 Stroke Collaborators. Global, regional, and national burden of stroke and its risk factors, 1990-2019: a systematic analysis for the global burden of disease study 2019. Lancet Neurol. 2021;20(10):795–820. https://doi.org/10.1016/S1474-4422(21)00252-0.
2. Abissegue G, Yakubu SI, Ajay AS, Niyi-Odumosu F. A systematic review of the epidemiology and the public health implications of stroke in sub-Saharan Africa. J Stroke Cerebrovasc Dis. 2024;33(8):107733. https://doi.org/10.1016/j.jstrokecerebrovasdis.2024.107733.
3. Langhorne P, O'Donnell MJ, Chin SL, Zhang H, Xavier D, et al. Practice patterns and outcomes after stroke across countries at different economic levels (INTERSTROKE): an international observational study. Lancet. 2018;391(10134):2019–27. https://doi.org/10.1016/S0140-6736(18)30802-X.
4. Krishnamurthi RV, Moran AE, Feigin VL, Barker-Collo S, Norrving B, et al. Stroke prevalence, mortality and disability-adjusted life years in adults aged 20-64 years in 1990-2013: data from the global burden of disease 2013 study. Neuroepidemiology. 2015;45(3):190–202. https://doi.org/10.1159/000441098.
5. Belagaje SR. Stroke rehabilitation. Continuum. 2017;23(1):238–53. https://doi.org/10.1212/CON.0000000000000423.
6. Hylin MJ, Kerr AL, Holden R. Understanding the mechanisms of recovery and/or compensation following injury. Neural Plast. 2017;2017:7125057. https://doi.org/10.1155/2017/7125057.
7. Dobkin BH. Strategies for stroke rehabilitation. Lancet Neurol. 2004;3(9):528–36. https://doi.org/10.1016/S1474-4422(04)00851-8.
8. Chen WC, Hsiao MY, Wang TG. Prognostic factors of functional outcome in post-acute stroke in the rehabilitation unit. J Formos Med Assoc. 2022;121(3):670–8. https://doi.org/10.1016/j.jfma.2021.07.009.
9. Langhammer B, Sunnerhagen KS, Lundgren-Nilsson Å, Sällström S, Becker F, et al. Factors enhancing activities of daily living after stroke in specialized rehabilitation: an observational multicenter study within the Sunnaas International Network. Eur J Phys Rehabil Med. 2017;53(5):725–34. https://doi.org/10.23736/S1973-9087.17.04489-6.

10. Feigin VL, Owolabi MO, World Stroke Organization–Lancet Neurology Commission Stroke Collaboration Group. Pragmatic solutions to reduce the global burden of stroke: a World Stroke Organization-Lancet Neurology Commission. Lancet Neurol. 2023;22(12):1160–206. https://doi.org/10.1016/S1474-4422(23)00277-6.
11. WHO. The international classification of functioning, disability and health. Geneva: World Health Organization; 2001.
12. WHO & World Bank. World report on disability. Geneva: WHO; 2011.
13. Talbot LR, Viscogliosi C, Desrosiers J, Vincent C, Rousseau J, et al. Identification of rehabilitation needs after a stroke: an exploratory study. Health Qual Life Outcomes. 2004;2:53.
14. Vincent C, Deaudelin I, Robichaud L, Rousseau J, Viscogliosi C, et al. Rehabilitation needs for older adults with stroke living at home: perceptions of four populations. BMC Geriatr. 2007;7:20.
15. Hackett ML, Yapa C, Parag V, Anderson CS. The frequency of depression after stroke: a systematic review of observational studies. Stroke. 2005;36:1330–40.
16. Feibel JH, Springer CJ. Depression and failure to resume social activities after stroke. Arch Phys Med Rehabil. 1982;63:276–8.
17. Hare R, Rogers H, Lester H, McManus R, Mant J. What do stroke patients and their carers want from community services? Fam Pract. 2006;23:131–6.
18. Moreland JD, Depaul VG, Dehueck AL, et al. Needs assessment of individuals with stroke after discharge from hospital stratified by acute functional Independence measure score. Disabil Rehabil. 2009;31(26):2185–95.
19. Burton CR. Living with stroke: a phenomenological study. J Adv Nurs. 2000;32:301–9.
20. King RB, Semik PE. Stroke caregiving: difficult times, resources use, and needs during the first 2 years. J Gerontol Nurs. 2006;32:37–44.
21. Langhorne P, Bernhardt J, Kwakkel G. Stroke rehabilitation. Lancet. 2011;377(9778):1693–702.
22. Langhorne P, Coupar F, Pollock A. Motor recovery after stroke: a systematic review. Lancet Neurol. 2009;8:741–54.
23. Evans A, Perez I, Harraf F, Melbourn A, Steadman J, et al. Can differences in management processes explain different outcomes between stroke unit and stroke-team care? Lancet. 2001;58(9293):1586–92.
24. Legg LA, Lewis SR, Schofield-Robinson OJ, Drummond A, Langhorne P. Occupational therapy for adults with problems in activities of daily living after stroke. Cochrane Database Syst Rev. 2017;7(7):CD003585. https://doi.org/10.1002/14651858.CD003585.pub3.
25. Kamalakannan S, Sethuraman L, Chockalingam M, Samuelkamaleshkumar S, Moorthy SD, Srinivasan MK, Ramakrishnan J, Muthuvel T, Mani K. Quality of occupational therapy research in India - a descriptive review. Indian J Occup Ther. 2022;54(3):91–5. https://doi.org/10.4103/ijoth.ijoth_59_22.
26. Hanger HC, Wilkinson TJ, Mears A. Stroke discharges from a rehabilitation unit: 1-year and 5-year domicile outcomes. Function is important. Intern Med J. 2010;40:45–51.
27. Stroke Unit Trialists' Collaboration. Organized inpatient (stroke unit) care for stroke. Cochrane Database Syst Rev. 2007;4:CD000197. https://doi.org/10.1002/14651858.CD000197.pub2.
28. Turner-Stokes L, Nair A, Sedki I, Disler PB, Wade DT. Multi-disciplinary rehabilitation for acquired brain injury in adults of working age. Cochrane Database Syst Rev. 2005;3:CD004170. https://doi.org/10.1002/14651858.CD004170.pub2.
29. Kim K, Kim YM, Kim EK. Correlation between the activities of daily living of stroke patients in a community setting and their quality of life. J Phys Ther Sci. 2014;26(3):417–9. https://doi.org/10.1589/jpts.26.417.
30. de Haan R, Aaronson N, Limburg M, Hewer RL, van Crevel H. Measuring quality of life in stroke. Stroke. 1993;24(2):320–7. https://doi.org/10.1161/01.str.24.2.320.
31. Shao C, Wang Y, Gou H, Chen T. The factors associated with the deterioration of activities of daily life in stroke patients: a retrospective cohort study. Top Stroke Rehabil. 2024;31(1):21–8. https://doi.org/10.1080/10749357.2023.2194095.
32. Kim HE, Cho KH. Factor analysis related to the change in activities of daily living performance of stroke patients. Biomed Res Int. 2023;2023:6147413. https://doi.org/10.1155/2023/6147413.

33. Grimby G, Andrén E, Daving Y, Wright B. Dependence and perceived difficulty in daily activities in community-living stroke survivors 2 years after stroke: a study of instrumental structures. Stroke. 1998;29(9):1843–9. https://doi.org/10.1161/01.str.29.9.1843.
34. Hsueh IP, Huang SL, Chen MH, Jush SD, Hsieh CL. Evaluation of stroke patients with the extended activities of daily living scale in Taiwan. Disabil Rehabil. 2000;22(11):495–500. https://doi.org/10.1080/096382800413989.
35. Ghaffari A, Rostami HR, Akbarfahimi M. Predictors of instrumental activities of daily living performance in patients with stroke. Occup Ther Int. 2021;2021:6675680. https://doi.org/10.1155/2021/6675680.
36. Shaw L, Bhattarai N, Cant R, Drummond A, Ford GA, et al. An extended stroke rehabilitation service for people who have had a stroke: the EXTRAS RCT. Health Technol Assess. 2020;24(24):1–202. https://doi.org/10.3310/hta24240.
37. Liu M, Peng W, Miyawaki CE, Mo C, Luo Y, Gong N. The bidirectional relationship between homebound status and falls among older adults: a longitudinal study. BMC Geriatr. 2022;22(1):923. https://doi.org/10.1186/s12877-022-03588-1.
38. Almeida AS, Paguia A, Neves AP. Nursing interventions to empower family caregivers to manage the risk of falling in older adults: a scoping review. Int J Environ Res Public Health. 2024;21(3):246. https://doi.org/10.3390/ijerph21030246.
39. Brown C, Joshi M, Watkins C. Life and leisure activities following stroke or transient ischaemic attack (TIA): an observational, multi-Centre, 6-month follow-up study. Int J Environ Res Public Health. 2022;19(21):13848. https://doi.org/10.3390/ijerph192113848.
40. Badaru UM, Ogwumike OO, Adeniyi AF, Olowe OO. Variation in functional independence among stroke survivors having fatigue and depression. Neurol Res Int. 2013;2013:842980. https://doi.org/10.1155/2013/842980.
41. Nelles G, Jentzen W, Jueptner M, Müller S, Diener HC. Arm training induced brain plasticity in stroke studied with serial positron emission tomography. NeuroImage. 2001;13(6):1146–54. https://doi.org/10.1006/nimg.2001.0757.
42. Maciejasz P, Eschweiler J, Gerlach-Hahn K, Jansen-Troy A, Leonhardt S. A survey on robotic devices for upper limb rehabilitation. J Neuroeng Rehabil. 2014;11:3. https://doi.org/10.1186/1743-0003-11-3.
43. van Ommeren AL, Smulders LC, Prange-Lasonder GB, Buurke JH, Veltink PH, Rietman JS. Assistive technology for the upper extremities after stroke: systematic review of users' needs. JMIR Rehabil Assist Technol. 2018;5(2):e10510. https://doi.org/10.2196/10510.
44. Chen PW, Baune NA, Zwir I, Wang J, Swamidass V, Wong AWK. Measuring activities of daily living in stroke patients with motion machine learning algorithms: a pilot study. Int J Environ Res Public Health. 2021;18(4):1634. https://doi.org/10.3390/ijerph18041634.
45. Willems EMG, Vermeulen J, van Haastregt JCM, Zijlstra GAR. Technologies to improve the participation of stroke patients in their home environment. Disabil Rehabil. 2022;44(23):7116–26. https://doi.org/10.1080/09638288.2021.1983041.
46. Gandhi DBC, Kamalakannan S, Urimubenshi G, Sebastian IA, Montanaro VVA, et al. GCSR collaborators. Stroke rehabilitation clinical practice guidelines in low- and middle-income countries: a systematic review of quality and unique features. Cerebrovasc Dis. 2024;2:1–12. https://doi.org/10.1159/000539999.
47. Kamalakannan S, Srivastava A, Chockalingam M, Gandhi D, Dhamija RK, Solomon J, Akkunje PS, Chitnis S, Someshwar H, Surya N. Health Systems for Rehabilitation - critical appraisal of the National Programme for stroke in India. J Indian Fed Neurorehabilit. 2024;1(1):18–26.
48. Kamalakannan S, Gudlavalleti Venkata M, Prost A, Natarajan S, Pant H, Chitalurri N, et al. Rehabilitation needs of stroke survivors after discharge from Hospital in India. Arch Phys Med Rehabil. 2016;97(9):1526–1532.e9. https://doi.org/10.1016/j.apmr.2016.02.008.
49. Sidek NN, Kamalakannan S, Tengku Ismail TA, Musa KI, Ibrahim KA, et al. Experiences and needs of the caregivers of stroke survivors in Malaysia-a phenomenological exploration. Front Neurol. 2022;13:996620. https://doi.org/10.3389/fneur.2022.996620.

50. Mairami FF, Warren N, Allotey PA, Reidpath DD. Contextual factors that shape recovery after stroke in Malaysia. Disabil Rehabil. 2020;42(22):3189–98. https://doi.org/10.1080/09638288.2019.1588399.
51. World Health Organization. Community-based rehabilitation: CBR guidelines. Geneva: World Health Organization; 2010.
52. Fearon P, Langhorne P. Early supported discharge Trialists. Services for reducing duration of hospital care for acute stroke patients. Cochrane Database Syst Rev. 2012;9:CD000443. https://doi.org/10.1002/14651858.CD000443.pub3.
53. Outpatient Service Trialists. Therapy-based rehabilitation services for stroke patients at home. Cochrane Database Syst Rev. 2003;1:CD002925. https://doi.org/10.1002/14651858.CD002925.pub2.
54. Van Niekerk SM, Inglis-Jassiem G, Kamalakannan S, Fernandes S, Webster J, English R, Smythe T, Louw QA. Achieving universal health coverage for people with stroke in South Africa: protocol for a scoping review. BMJ Open. 2020;10(10):e041221. https://doi.org/10.1136/bmjopen-2020-041221.

Post-Stroke Unaddressed Complications (Fatigue, Incontinence, Depression, and Pain)

8

Pradeepa Nayak, Rinita Mascarenhas, Vinicius Montanaro, and Aradhana Shukla

Abstract

Post-stroke complications such as fatigue, incontinence, depression, and pain significantly affect recovery post-stroke and quality of life. Post-stroke fatigue (PSF) is a common issue, affecting nearly half of stroke survivors. Unlike regular fatigue, PSF is often persistent and does not improve with rest, leading to physical, emotional, and cognitive exhaustion. PSF is assessed using tools like the Fatigue Severity Scale, and management strategies include medications, tailored exercise programmes, energy-saving techniques, and complementary therapies such as mindfulness and yoga. Incontinence, particularly urinary incontinence (UI), affects 38%–60% of stroke survivors shortly after the event, with about 19% experiencing long-term symptoms. Management often involves tailored interventions like timed voiding, pelvic floor exercises, and occasionally, electri-

P. Nayak (✉)
Department of Health Professions, Manchester Metropolitan University, Birley Fields Campus, Manchester, UK
e-mail: p.nayak@mmu.ac.uk

R. Mascarenhas
Department of Neurology, Christian Medical College and Hospital, Ludhiana, Punjab, India

V. Montanaro
Neurological Rehabilitation SARAH Network of Rehabilitation Hospitals, Brasília, Brazil

Radbou University Medical Center, Nijmegen, Netherlands

Programa de Pós Grduação em Neurologia da Universidade Federal Fluminense, Niterói, Brazil

SARAH Network of Rehabilitation Hospitals, Brasília, Brazil

A. Shukla
Department of Physical Medicine and Rehabilitation, King George's Medical University, Lucknow, Uttar Pradesh, India

D. B. C. Gandhi, J. D Pandian (eds.), *Stroke Rehabilitation in Low-Resource Settings*, https://doi.org/10.1007/978-981-92-0403-8_8

cal stimulation. Faecal incontinence, also common post-stroke, is addressed through dietary changes, medications, and emerging techniques like transanal irrigation. Post-stroke depression (PSD) is another significant complication, affecting up to one-third of stroke survivors. It hampers recovery, reduces participation in rehabilitation, and lowers overall quality of life. In low-resource settings, tools like the Patient Health Questionnaire (PHQ-9) and telemedicine solutions enable early detection. Treatment involves a combination of medications, simplified cognitive-behavioural therapy, physical activity programmes, and community-driven social support. Pain after stroke is a common issue but remains relatively unaddressed due to an inadequate understanding of pathophysiology and clinical consequences. Pharmacological interventions like tricyclic antidepressants (TCAs), gabapentin, botulinum toxin, and non-pharmacological interventions such as mirror therapy and functional electrical stimulation can be utilized to treat post-stroke pain.

Keywords

Post-stroke fatigue · Post-stroke depression · Post-stroke incontinence · Stroke rehabilitation

This chapter focusses on understanding and managing important post-stroke complications—fatigue, incontinence, depression, and pain by discussing their causes, assessment, and treatment approaches. An overview of the causes of these complications is presented in Figure 8.1. By integrating medical, psychological, and community-based interventions, this chapter underscores the importance of personalized care strategies to enhance recovery and quality of life post-stroke.

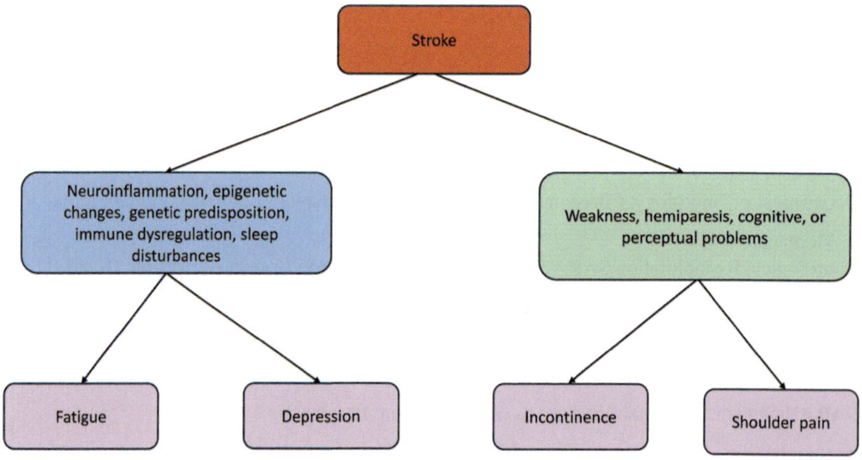

Fig. 8.1 Overview of causes of post-stroke complications

8.1 Post-Stroke Fatigue

Fatigue is one of the most common complications of stroke. Almost one in two stroke survivors experiences post-stroke fatigue [1]. Fatigue after stroke is different than normal fatigue. Post-stroke fatigue (PSF) can be described as an overwhelming sense of tiredness, lack of energy, and feeling of exhaustion. Stroke Recovery and Rehabilitation Roundtable consensus statement defines PSF as "feeling of exhaustion, weariness or lack of energy that can be overwhelming, and which can involve physical, emotional, cognitive and perceptual contributors, which is not relieved by rest and affects a person's daily life" [2]. In contrast to typical fatigue, this type of fatigue is unrelated to level of activity or busyness and does not always improve with rest. Stroke survivors might report feeling weak, worn out, drowsy, or unable to motivate themselves. For some, it could exacerbate symptoms like weakness, amnesia, distractibility, and slurred speech that are related to the stroke. PSF negatively impacts stroke survivors' participation in activities of daily living and the quality of life [3, 4]. PSF can lead to social isolation [5] and affect community reintegration post-stroke [6].

8.1.1 Assessment

PSF is an experience which is unique to the individual. Due to its subjective nature and lack of standardized diagnostic criteria, it is difficult to measure PSF, though various tools can be used clinically to monitor fatigue. Due to the high prevalence of the condition, everyone who has had a stroke should be screened for the presence of fatigue. Scales which are useful for screening and monitoring PSF include Fatigue Severity Scale (FSS) [7], Fatigue Impact Scale [8], Modified Fatigue Impact Scale (MFIS), Stroke Fatigue Clinical Assessment Tool (SF-CAT), Fatigue Scale for Motor and Cognitive (FSMC) functions [9], and Multidimensional Fatigue Symptom Inventory.

The 6-min walk test could be utilized to measure the fatigability objectively. Electromyography is a promising tool to assess muscle fatigue in stroke. The changes in electroencephalography can be investigated whilst performing simulated tasks to assess fatigue [10].

8.1.2 Management

Given the complexities of PSF, effective management necessitates a multidisciplinary and individualistic approach. A combination of medical, psychological, and lifestyle interventions can be used to effectively manage PSF.

Medical interventions: Dopamine reuptake inhibitors, methylphenidate, and modafinil have been used to treat PSF [11, 12]. Methylphenidate reduces the cost of mental labour (i.e., perceived effort). Research showed that intake of methylphenidate was associated with higher engagement in challenging cognitive tasks compared with the control group [13]. Along with this, managing underlying conditions such as sleep disorders, and treating anxiety and depression, can help alleviate PSF.

Physical rehabilitation: Regular, moderate exercise has been shown to reduce fatigue and improve energy levels in stroke survivors [14, 15]. Physical activity programmes should be tailored to the individual's abilities and may include aerobic exercises, strength training, and balance exercises. Physical therapy can also help rebuild stamina and resilience.

Transcranial direct current stimulation: There is mixed evidence on the effectiveness of non-invasive brain stimulation in PSF [16]. Preliminary research has shown that a single session of anodal transcranial direct current stimulation has the potential to reduce PSF and the improvements were maintained up to a week post stimulation [17]. Randomized controlled trials using anodal direct current stimulation [18] and bilateral transcranial direct current stimulation [19] have demonstrated reduction in fatigue symptoms during the course of the intervention; however, the improvements were not maintained after the end of the intervention.

Psychoeducation: Psychoeducation and cognitive behavioural therapy can be used to address mental and emotional aspects of fatigue [20]. These interventions help in alleviating negative thought patterns, enhance motivation, reducing stress and anxiety, and coping with PSF.

Energy conservation techniques: Energy conservation techniques, such as activity planning, task prioritisation, and scheduled rest breaks, can help manage post-stroke fatigue more effectively by reducing physical and cognitive overload [21].

Complementary Therapies: Complementary therapies such as mindfulness meditation, yoga, and acupuncture may help relieve PSF by lowering stress and improving energy balance. Prescription of traditional Chinese medicines [22] and vitamin B12 supplementation [23] have also been shown to be effective in managing PSF.

Fostering reasonable expectations and encouraging adherence to treatment programmes requires educating patients and caregivers about PSF. Being aware of the illness may alleviate frustration or a sense of helplessness.

8.2 Incontinence

Dysfunction of the bladder and/or bowel may be caused by a combination of stroke-related impairments (e.g., weakness, cognitive, or perceptual impairments).

8.2.1 Urinary Incontinence

Post-stroke urinary incontinence (UI) is a common complication, affecting between 38% and 60% of stroke survivors within the first month. While many cases resolve spontaneously within 8 weeks, approximately 19% of patients may experience persistent UI 6 months or later [24].

Three types of UI disorders can be addressed through occupational and/or physical therapy:

- Urge UI (UUI), characterized by involuntary leakage accompanied or preceded by urgency, can result from stroke-related infarcts or cerebral oedema impacting central micturition pathways.
- Functional UI (FUI) arises from functional limitations, such as hemiparesis or reduced mobility, or environmental barriers like inaccessible toilets.
- Stress UI (SUI), the involuntary leakage associated with effort, sneezing, or coughing, often predates stroke but may worsen post-stroke due to coughing from dysphagia or aspiration.

While no specific UI assessment exists for stroke patients, the International Consultation on Incontinence recommends a protocol that includes a urinary diary, symptom and quality-of-life questionnaires, a physical exam with pelvic floor evaluation, an assessment of toilet transfers, and consideration of adaptive equipment. Additionally, stroke-specific and general functional measures (functional independence measure—FIMS) include items that assess toileting needs [25].

8.2.2 Rehabilitation

Based on recommendations from three Clinical Practice Guidelines, there is consensus opinion that timed voiding or prompted voiding should be implemented for urinary retention or incontinence in cooperative and mobile individuals with stroke. Timed voiding is a toileting assistance programme based on fixed time intervals, commonly used to manage urinary incontinence (UI) in individuals unable to toilet independently. Prompted voiding, on the other hand, is a behavioural intervention designed to encourage individuals, with or without cognitive impairments, to initiate their own toileting [26].

The evidence regarding pelvic floor exercises among stroke survivors is sparse. Yet coordinating diaphragmatic breathing with pelvic floor muscle activity can enhance control over pelvic floor relaxation and improve oxygenation to the pelvic area. In this technique, clients learn to engage their pelvic floor muscles during exhalation (eccentric movement of the diaphragm, concentric activity of the pelvic floor muscles) and relax these muscles on inhalation. This exercise can be practiced in various positions, such as supine, side-lying, on all fours, seated, or standing. Further pelvic floor exercises with functional movements—such as contracting during exhalation when standing up from a chair—can develop these responses into automatic patterns through repetition and procedural learning.

There is ambiguity with the limited and varied methodology published trials that assess non-implanted electrical stimulation for post-stroke incontinence. Current evidence indicates that early, frequent electrical stimulation may be more effective in treating incontinence.

8.2.3 Faecal Incontinence

Faecal incontinence (FI) is a frequent complication following acute stroke, with prevalence rates ranging from 23% to 60% in the acute phase. Bowel dysfunction symptoms include constipation and diarrhoea, with limited toilet access and use of constipating medications as two modifiable post-stroke risk factors [27]. This condition adversely impacts the quality of life for both patients and their caregivers, often restricting social activities. Despite its significance, management strategies largely rely on clinical experience due to limited evidence on effective interventions [28].

FI can be treated along the lines of dietary adjustments, anal plugs, transanal irrigation, and surgery. The evidence regarding the use of electrical stimulation is limited but has been found to be effective [29].

8.3 Post-Stroke Depression

Post-stroke depression (PSD) is a common yet often underdiagnosed complication that profoundly affects recovery and quality of life in stroke survivors. Estimated to affect up to one-third of all stroke patients, PSD can lead to poorer functional outcomes, reduced adherence to rehabilitation, and increased mortality [30].

8.3.1 Assessment

Effective evaluation of PSD relies on validated screening tools that are both accessible and accurate for stroke survivors with varying levels of cognitive and communication abilities. The Patient Health Questionnaire (PHQ-9) and the Beck Depression Inventory (BDI-II) are commonly used tools for assessing depressive symptoms in stroke patients. Recent studies highlight that these tools can be administered by trained non-specialists with good reliability, making them viable in LMICs where specialist availability is limited [31]. The PHQ-9, in particular, has shown efficacy when administered by primary care providers or community health workers (CHWs) who receive minimal training, a feasible solution in LMICs [32].

8.3.2 Management

Effective treatment of PSD in LMICs encompasses pharmacological, psychological, and social support interventions adapted to local resource availability.

1. Pharmacological treatment.

   Selective serotonin reuptake inhibitors (SSRIs), such as fluoxetine and sertraline, are the most researched antidepressants for PSD, demonstrating efficacy in reducing depressive symptoms and even improving neurological recovery in some cases [33]. Given the limited availability of SSRIs in many LMICs, partnerships with NGOs and government programmes have been essential in ensuring these medications reach underserved areas. However, monitoring for side effects, particularly bleeding risks associated with SSRIs, is challenging in low-resource settings. As such, community health workers are often trained to observe and report any adverse effects, enabling a timely response despite limited professional oversight [34].

2. Psychological interventions.

   Cognitive behavioural therapy (CBT) has shown significant efficacy in treating PSD but requires trained mental health professionals, who are often scarce in LMICs. An effective alternative in these settings is "task-sharing," where trained community health workers deliver simplified CBT, focussing on strategies to manage depressive thoughts and promote adaptive behaviours [35]. Research in India and Nepal demonstrates that CHWs can deliver brief, CBT-based interventions with success in reducing depressive symptoms among stroke survivors [36]. Group-based therapies led by CHWs also show potential in improving social connectedness and reducing depressive symptoms, particularly in community-centred cultures.

3. Social support interventions.

   Community-based programmes that integrate family and peer support have proven beneficial for PSD management. Social isolation is a risk factor for PSD, and creating peer support groups allows survivors to share experiences, learn adaptive coping mechanisms, and build a sense of community. In Tanzania, a pilot programme integrated PSD management into stroke clubs, with participants reporting a marked decrease in depressive symptoms through group support, family engagement, and structured social activities [37]. This community-oriented approach addresses PSD without high dependency on individual mental health resources.

4. Physical exercise and lifestyle modifications.

   Physical exercise is increasingly recognized as a beneficial intervention for PSD, as it enhances mood, reduces anxiety, and can improve overall physical recovery. In LMICs, where formal rehabilitation gyms may be unavailable, physical therapy routines often incorporate local resources, such as walking exercises in community areas or simple stretching routines. A recent study in Kenya found that structured walking groups for stroke survivors improved both physical outcomes and mood, underscoring the role of low-cost, exercise-based community programmes in addressing PSD [38].

8.4 Pain Conditions in Post-Stroke Phase

Pain after stroke is a common issue but relatively unaddressed due to inadequate understanding of pathophysiology and clinical consequences, yet it is reported to have a negative effect on the quality of life [39–41]. Through a broad search in the literature, pain in the post-stroke phase was mainly found to be central post-stroke pain, complex regional pain syndrome, shoulder-hand syndrome/hemiplegic shoulder, and pain associated with spasticity [42].

8.4.1 Central Post-Stroke Pain (CPSP)

First described by Dejerine and Roussy in 1906, as pain on the hemiplegic side, paroxysmal, often intolerable, not responding to any analgesics in patients with stroke [43]. The term CPSP is now described as neuropathic pain after stroke because researchers have shown the cause to be a lesion or dysfunction of the CNS (central pain of thalamic and extra-thalamic origin) [44]. CPSP, being a diagnosis of exclusion, presents with a scarcity of evidence-based treatment or recommendations for prevention. TCA, gabapentin/pregabalin could be first-line drugs, and failing one can try SSRI, lamotrigine, opioids [45]. Coping strategies and behavioural therapy might benefit, and rehabilitation is crucial in such conditions.

8.4.2 Complex Regional Pain Syndrome (CRPS)

Also known as reflex sympathetic dystrophy, it is characterized by burning pain, increased sensitivity to tactile stimulation, and vasomotor changes, including oedema and changes in skin temperature and colour, which might be a result of local hyperactivity of the sympathetic nervous system with or without sensitization of the somatic sensory pathway, overactivation of inflammatory responses, and hypoxia [46]. CRPS in stroke earlier was managed by sympathetic blockade, showing minimal efficacy [47]. Recently, prophylactic calcitonin administration showed a significant reduction in the onset of CRPS [48]. Pharmacological trials in post-stroke CRPS are lacking greatly. Non-pharmacological therapies that may help include mirror therapy.

8.4.3 Hemiplegic Shoulder/Shoulder-Hand Syndrome

Hemiplegic shoulder broadly encompasses subluxation, muscle imbalance and weakness, adhesive capsulitis, spasticity as well as CRPS [49]. The extra stress from external forces on a hemiplegic limb leads to subluxation, which is one of the most common painful complications post-stroke specific to the shoulder joint [42].

Previous evidence in treating this condition have evolved from cuff slings, Bobath rolls, and shoulder strapping to functional electrical stimulation in early stages to prevent subluxation [50] and peripheral nerve stimulations to reduce pain [51]. However current evidence states the need for a proactive and multifaceted approach. Early initiation of exercises targeting the shoulder stabilizers is essential. Proper positioning plays a critical role in avoiding microtrauma and shoulder capsule stretching; the affected shoulder should be supported during sitting, standing, and walking with the use of slings, lap trays, or pillows. However, hand splints are not recommended; such measures help maintain joint integrity and reduce the risk of further injury during daily activities and ambulation. Although the evidence for the use of electrical stimulation and taping remains limited, these modalities may be considered as adjuncts, as their use is unlikely to cause harm.

8.4.4 Pain Associated with Spasticity

Spasticity is an involuntary contraction of muscle groups from an exaggeration of the stretch reflex, often painful. Consequences of poor management may lead to contracture formation, spastic, immobile limbs with and an increased predisposition for skin breakdowns [42].

The only objective of spasticity treatment is to maintain the range of motion of the affected muscle groups. Conditions not managed conservatively can resort to antispastic medications, e.g., Baclofen, tizanidine, dantrolene, diazepam, etc. Baclofen can be delivered intrathecally as well. Botulinum neurotoxin has been amply studied with minimal side effects for reducing pain associated with spasticity. Also, nerve blocks, motor point blocks using phenol, or alcohol can help in reducing pain from spasticity. Other treatments which need more evidence include therapeutic ultrasound, transcutaneous electrical stimulations, and rTMs. Table 8.1 summarises the key post-stroke complications discussed above (Table 8.1).

Table 8.1 Summary of post-stroke complications

Complication	Prevalence	Clinical features	Risk factors	Assessment	Management
Post-stroke fatigue	40–50%	Feeling of exhaustion weariness or lack of energy, not relieved by rest	Female sex, haemorrhagic stroke, damage to thalamus, genetic factors, reduced physical fitness, sleep disturbances, anxiety, and depression	Fatigue severity scale, fatigue impact scale, modified fatigue impact scale, stroke fatigue clinical assessment tool	Dopamine reuptake inhibitors, regular moderate exercises, transcranial direct current stimulation, psychoeducation, energy conservation techniques, mindfulness meditation, and yoga
Incontinence	38–60%	Changes in urgency, frequency and control of bowel and bladder	Damaged neural pathways, and motor, cognitive impairments	Urinary diary, symptom and quality-of-life questionnaires, pelvic floor evaluation	Timed voiding, prompted voiding, coordination of breathing with pelvic floor muscle exercises, and electrical stimulation
Depression	20–65%	Low mood, lack of sleep, lack of motivation	History of depression, social isolation, poor functional status	Patient health questionnaire (PHQ-9) and the Beck depression inventory (BDI-II)	Selective serotonin reuptake inhibitors, cognitive behavioural therapy, social support, and physical exercises
Shoulder pain	30–60%	Pain with movement, spasticity, subluxation	Hemiparesis, immobility, subluxation, severe spasticity	Visual analogue scale, numeric rating scale, Fugl–Meyer assessment for upper extremity, disabilities of the arm, shoulder and hand (DASH)	Shoulder steroid injections, botulinum toxin A, positioning, electrical stimulation

8.4.5 Current Picture

What Does Evidence Say for Various Secondary Complications

- Shoulder injections (e.g., subacromial steroids for rotator cuff syndrome or suprascapular nerve block with methylprednisolone and bupivacaine) may be used to reduce pain.
- Botulinum Toxin A may be used for shoulder pain associated with upper limb spasticity.
- Electrical stimulation may be used within the first 6 months post-stroke to prevent or reduce shoulder subluxation.
- Shoulder strapping is not recommended for preventing or managing subluxation.
- All stroke survivors with suspected urinary issues should undergo a structured assessment by trained personnel, supported by portable bladder ultrasound where appropriate.
- Hospitalized patients with confirmed continence problems should have a documented and monitored continence management plan, with reassessment and specialist referral if issues persist after discharge.
- For urge incontinence, management may include anticholinergic medications, bladder retraining (prompted/scheduled voiding), and use of containment aids if continence cannot be achieved.
- Suspected faecal continence issues should be assessed through structured evaluation, including rectal examination where needed.
- Management should address constipation, faecal overflow, or incontinence, along with patient education.
- Bowel retraining strategies, including dietary adjustments, timing of intake, and use of the gastro-colic reflex, should be applied.
- Emotionalism may be treated with antidepressants such as SSRIs or tricyclic antidepressants.
- Antidepressants may also be used to prevent and treat post-stroke depression, with choice tailored to individual risk–benefit profiles.
- Psychological therapies (e.g., problem-solving therapy, motivational interviewing) may be used for both prevention and treatment of depression.
- Structured exercise programmes, particularly high-intensity or resistance training, may help improve depressive symptoms.
- Additional options for depression may include non-invasive brain stimulation (e.g., rTMS) and acupuncture.
- All stroke survivors should be assessed for fatigue using the stroke fatigue clinical assessment tool (SF-CAT).
- When measuring fatigue in studies or assessments, tools such as the Fatigue Severity Scale (FSS-7), visual analogue scales (VAS), and qualitative measures should be used, ensuring they are accessible for individuals with aphasia.

Author Contributions The authors contributed to different aspects of the chapter on post-stroke complications. Pradeepa Nayak focussed on post-stroke fatigue, discussing its prevalence, assessment methods, and management strategies. Rinita Mascarenhas addressed incontinence, detailing interventions for urinary and faecal incontinence, including pelvic floor exercises. Vinícius Montanaro focussed on post-stroke depression, emphasizing its impact on recovery, tools for early detection, and treatments such as cognitive-behavioural therapy and social support. Aradhana Shukla examined post-stroke pain, highlighting key conditions and management techniques like pharmacological therapies and non-pharmacological interventions.

References

1. Alghamdi I, Ariti C, Williams A, Wood E, Hewitt J. Prevalence of fatigue after stroke: a systematic review and meta-analysis. Eur Stroke J. 2021;6(4):319–32. https://doi.org/10.1177/23969873211047681.
2. English C, Simpson DB, Billinger SA, Churilov L, Coupland KG, Drummond A, Kuppuswamy A, Kutlubaev MA, Lerdal A, Mahmood A, Moseley GL, Pittman QJ, Riley EA, Sutherland BA, Wong CH, Corbett D, Mead G. A roadmap for research in post-stroke fatigue: consensus-based core recommendations from the third stroke recovery and rehabilitation roundtable. Int J Stroke. 2024;19(2):133–44. https://doi.org/10.1177/17474930231189135.
3. Aali G, Drummond A, das Nair R, Shokraneh F. Post-stroke fatigue: a scoping review. F1000Res. 2020;9:242. https://doi.org/10.12688/f1000research.22880.2.
4. Lerdal A, Gay CL. Acute-phase fatigue predicts limitations with activities of daily living 18 months after first-ever stroke. J Stroke Cerebrovasc Dis. 2017;26(3):523–31. https://doi.org/10.1016/j.jstrokecerebrovasdis.2016.11.130.
5. Hinkle JL, Becker KJ, Kim JS, Choi-Kwon S, Saban KL, McNair N, Mead GE, American Heart Association Council on Cardiovascular and Stroke Nursing and Stroke Council. Poststroke fatigue: emerging evidence and approaches to management: a scientific statement for healthcare professionals from the American Heart Association. Stroke. 2017;48(7):e159–70. https://doi.org/10.1161/STR.0000000000000132.
6. Ho LYW, Lai CYY, Lai CKY, Ng SSM. Fatigue predicts level of community integration in people with stroke. Top Stroke Rehabil. 2024;31(5):464–73. https://doi.org/10.1080/10749357.2023.2298536.
7. Ozyemisci-Taskiran O, Batur EB, Yuksel S, Cengiz M, Karatas GK. Validity and reliability of fatigue severity scale in stroke. Top Stroke Rehabil. 2019;26(2):122–7. https://doi.org/10.1080/10749357.2018.1550957.
8. Batur EB, Ozyemisçi-Taskiran O, Yuksel S, Cengiz M, Karatas GK. Validity and reliability of the fatigue impact scale in stroke. Top Stroke Rehabil. 2022;29(7):526–37. https://doi.org/10.1080/10749357.2021.1978629.
9. Hubacher M, Calabrese P, Bassetti C, Carota A, Stöcklin M, Penner IK. Assessment of post-stroke fatigue: the fatigue scale for motor and cognitive functions. Eur Neurol. 2012;67(6):377–84. https://doi.org/10.1159/000336736.
10. Chen W, Jiang T, Huang H, Zeng J. Post-stroke fatigue: a review of development, prevalence, predisposing factors, measurements, and treatments. Front Neurol. 2023;14:1298915. https://doi.org/10.3389/fneur.2023.1298915.
11. Bivard A, Lillicrap T, Krishnamurthy V, Holliday E, Attia J, Pagram H, Nilsson M, Parsons M, Levi CR. MIDAS (modafinil in debilitating fatigue after stroke): a randomized, double-blind, placebo-controlled, cross-over trial. Stroke. 2017;48(5):1293–8. https://doi.org/10.1161/STROKEAHA.116.016293.
12. Blockmans D, Persoons P, Van Houdenhove B, Bobbaers H. Does methylphenidate reduce the symptoms of chronic fatigue syndrome? Am J Med. 2006;119(2):167.
13. Hofmans L, Papadopetraki D, van den Bosch R, Määttä JI, Froböse MI, Zandbelt BB, Westbrook A, Verkes RJ, Cools R. Methylphenidate boosts choices of mental labor over leisure depend-

ing on striatal dopamine synthesis capacity. Neuropsychopharmacology. 2020;45(13):2170–9. https://doi.org/10.1038/s41386-020-00834-1.
14. Saunders DH, Sanderson M, Hayes S, Johnson L, Kramer S, Carter DD, Jarvis H, Brazzelli M, Mead GE. Physical fitness training for stroke patients. Cochrane Database Syst Rev. 2020;3(3):CD003316. https://doi.org/10.1002/14651858.CD003316.pub7.
15. Tai D, Falck RS, Davis JC, Vint Z, Liu-Ambrose T. Can exercise training promote better sleep and reduced fatigue in people with chronic stroke? A systematic review. J Sleep Res. 2022;31(6):e13675. https://doi.org/10.1111/jsr.13675.
16. Kuppuswamy A, Billinger S, Coupland KG, English C, Kutlubaev MA, Moseley L, Pittman QJ, Simpson DB, Sutherland BA, Wong C, Corbett D. Mechanisms of post-stroke fatigue: a follow-up from the third stroke recovery and rehabilitation roundtable. Neurorehabil Neural Repair. 2024;38(1):52–61. https://doi.org/10.1177/15459683231219266.
17. De Doncker W, Ondobaka S, Kuppuswamy A. Effect of transcranial direct current stimulation on post-stroke fatigue. J Neurol. 2021;268:2831–42.
18. Dong XL, Sun X, Sun WM, Yuan Q, Yu GH, Shuai L, Yuan YF. A randomized controlled trial to explore the efficacy and safety of transcranial direct current stimulation on patients with post-stroke fatigue. Medicine (Baltimore). 2021;100(41):e27504. https://doi.org/10.1097/MD.0000000000027504. PMID: 34731132; PMCID: PMC8519229.
19. Ulrichsen KM, Kolskår KK, Richard G, Pedersen ML, Alnaes D, Dørum ES, Sanders AM, Tornås S, Maglanoc LA, Engvig A, Ihle-Hansen H, Nordvik JE, Westlye LT. No add-on effect of tDCS on fatigue and depression in chronic stroke patients: a randomized sham-controlled trial combining tDCS with computerized cognitive training. Brain Behav. 2022;12(7):e2643. https://doi.org/10.1002/brb3.2643.
20. Nguyen S, Wong D, McKay A, Rajaratnam SMW, Spitz G, Williams G, Mansfield D, Ponsford JL. Cognitive behavioural therapy for post-stroke fatigue and sleep disturbance: a pilot randomised controlled trial with blind assessment. Neuropsychol Rehabil. 2019;29(5):723–38. https://doi.org/10.1080/09602011.2017.1326945.
21. Acciarresi M, Bogousslavsky J, Paciaroni M. Post-stroke fatigue: Epidemiology, clinical characteristics and treatment. European Neurology, 2014;72(5–6):255–261. https://doi.org/10.1159/000363763.
22. Guo C, Wang Y, Wang S, Zhang S, Tai X. Effect and mechanism of traditional Chinese medicine exercise therapy on stroke recovery. Evid Based Complement Alternat Med. 2023;2023(1):5507186.
23. Gurak SV, Parfenov VA. Asthenia after stroke and myocardial infarction and its treatment with Enerion. Clin Gerontol. 2005;8:9–12.
24. Dumoulin C, Korner-Bitensky N, Tannenbaum C. Urinary incontinence after stroke: identification, assessment, and intervention by rehabilitation professionals in Canada. Stroke. 2007;38(10):2745–51.
25. Gordon C, Davidson CE, Roffe C, Clegg A, Booth J, Lightbody CE, Watkins C. Evaluating methods of detecting and determining the type of urinary incontinence in adults after stroke: a systematic review. Neurourol Urodyn. 2024;43(2):364–81.
26. Özden F, Tümtürk İ, Özkeskin M, Bakırhan S. The effect of pelvic floor muscle training on urinary incontinence in patients with stroke: a systematic review and meta-analysis. Ir J Med Sci. 2023;192(3):1481–95.
27. Australian and New Zealand Living Clinical Guidelines for Stroke Management. Chapter 6 of 8: Managing complications||Faecal incontinence.
28. Lazaro R, Umphred D. Umphred's neurorehabilitation for the physical therapist assistant. New York: Taylor & Francis; 2024.
29. Cruz E, Miller C, Zhang W, Rogers K, Lee HJ, Wells Y, Lannin NA. Does non-implanted electrical stimulation reduce post-stroke urinary or fecal incontinence? A systematic review with meta-analysis. Int J Stroke. 2022;17(4):378–88.
30. Towfighi A, Ovbiagele B, El Husseini N, Horn N. Post-stroke depression: a scientific statement for healthcare professionals. Stroke. 2017;48(2):e30–43.

31. Olver JS, Hopwood M. Screening for depression in post-stroke patients. J Stroke Med. 2020;3(4):123–30.
32. Mahmud A, Raghavan P, Thomas D. Validation of the PHQ-9 for depression screening by community health workers in LMICs. Global Mental Health J. 2021;7(1):12–8.
33. Villa M, Moriguchi L. Selective serotonin reuptake inhibitors in post-stroke depression management. Eur J Neurol. 2021;28(1):73–80.
34. Oyebode T, Okorie P. Observational strategies for SSRI side effects in rural health settings. Afr J Psychiatry. 2020;17(3):94–8.
35. Singla DR, Kohrt BA. Task-sharing in LMICs: CBT by non-specialists. Psychol Bull. 2017;143(3):197–206.
36. Chowdhary N, Patel V. Brief CBT interventions by community health workers in India. Global Mental Health. 2021;5(4):34–43.
37. Mosha J, Mahenge S. Stroke clubs and depression management: a Tanzanian model. Int J Stroke. 2023;18(2):102–8.
38. Njoroge R, Nyaga M. Community exercise programs for post-stroke patients in Kenya. Afr J Phys Med. 2022;15(3):154–61.
39. Jönsson AC, Lindgren I, Hallström B, Norrving B, Lindgren A. Prevalence and intensity of pain after stroke: a population based study focusing on patients' perspectives. J Neurol Neurosurg Psychiatry. 2006;77(5):590–5.
40. Kong KH, Woon VC, Yang SY. Prevalence of chronic pain and its impact on health-related quality of life in stroke survivors. Arch Phys Med Rehabil. 2004;85(1):35–40.
41. Widar M, Ahlström G, Ek AC. Health-related quality of life in persons with long-term pain after a stroke. J Clin Nurs. 2004;13(4):497–505.
42. Treister AK, Hatch MN, Cramer SC, Chang EY. Demystifying poststroke pain: from etiology to treatment. PM&R. 2017;9(1):63–75.
43. Wilkins RH, Brody IA. The thalamic syndrome. Arch Neurol. 1969;20(5):559.
44. Riddoch G. The clinical features of central pain. Lancet. 1938;231(5986):1150–6.
45. Klit H, Finnerup NB, Jensen TS. Central post-stroke pain: clinical characteristics, pathophysiology, and management. Lancet Neurol. 2009;8(9):857–68.
46. Chae J. Poststroke complex regional pain syndrome. Top Stroke Rehabil. 2010;17(3):151–62.
47. Cepeda MS, Lau J, Carr DB. Defining the therapeutic role of local anesthetic sympathetic blockade in complex regional pain syndrome: a narrative and systematic review. Clin J Pain. 2002;18(4):216–33.
48. Matayoshi S, Shimodozono M, Hirata Y, Ueda T, Horio S, Kawahira K. Use of calcitonin to prevent complex regional pain syndrome type I in severe hemiplegic patients after stroke. Disabil Rehabil. 2009;31(21):1773–9.
49. Chae J, Ng A, Yu DT, Kirsteins A, Elovic EP, Flanagan SR, Fang ZP. Intramuscular electrical stimulation for shoulder pain in hemiplegia: does time from stroke onset predict treatment success? Neurorehabil Neural Repair. 2007b;21(6):561–7.
50. Ada L, Foongchomcheay A. Efficacy of electrical stimulation in preventing or reducing subluxation of the shoulder after stroke: a meta-analysis. Austr J Physiother. 2002;48(4):257–67.
51. Chae J, Mascarenhas D, David TY, Kirsteins A, Elovic EP, Flanagan SR, Fang ZP. Poststroke shoulder pain: its relationship to motor impairment, activity limitation, and quality of life. Arch Phys Med Rehabil. 2007a;88(3):298–301.

Technology-Enhanced Post-Stroke Rehabilitation: What's Achievable in Low-Resource Settings

9

A. Srivastava, S. Balasubramanium, J. Baggio, N. S. Chawla, Aakhil Nabuhan, and Dorcas B. C. Gandhi

Abstract

Stroke is a leading cause of disability worldwide, leaving many survivors with persistent impairments requiring long-term rehabilitation. Access to adequate care is particularly challenging in low-resource settings, where affordability and availability of rehabilitation services remain major barriers. Recent advancements in technology offer promising solutions to enhance post-stroke recovery, addressing motor, sensory, cognitive, and emotional deficits. Despite their potential, the high cost of many technologies limits their integration in low-resource contexts. However, locally manufactured, affordable solutions demonstrate the feasibility of scaling these interventions. Further innovation and evidence-based implementation of affordable, context-specific technologies are critical for meeting the growing demand for stroke rehabilitation in underserved areas. Collaboration among researchers, healthcare professionals, and policymakers is

A. Srivastava (✉) · A. Nabuhan
Centre of Physical and Medical Rehabilitation, Kokilaben Dhiribhai Ambani, Mumbai, India

S. Balasubramanium
Department of Bioengineering, Christian Medical College, Vellore, India

J. Baggio
Medical School of Federal University of Alagoas, Arapiraca, Brazil

N. S. Chawla
Department of Physiotherapy, Manipal College of Health Professions, Manipal Academy of Higher Education (MAHE), Manipal, Karnataka, India

D. B. C. Gandhi
Department of Neurology, Christian Medical College & Hospital Ludhiana, Ludhiana, Punjab, India

Department of Physiotherapy, Manipal Academy of Higher Education, Manipal, Karnataka, India

D. B. C. Gandhi, J. D Pandian (eds.), *Stroke Rehabilitation in Low-Resource Settings*, https://doi.org/10.1007/978-981-92-0403-8_9

essential to ensure equitable access to these advancements, thereby improving patient outcomes and quality of life globally.

Keywords

Stroke rehabilitation • Low-resource settings • Affordable technology • Robotics and virtual reality • Non-motor deficits

Learning Objectives

By the end of this chapter, readers will be able to:

The use of technology in stroke rehabilitation is increasingly recognized as essential due to its potential to enhance recovery outcomes. Technology offers immense potential to address the multifaceted nature of post-stroke disabilities. This chapter highlights the key technological innovations in stroke rehabilitation for different motor and non-motor domains. Robotic devices can aid motor recovery, while virtual reality-based platforms create immersive environments that engage multiple sensory and cognitive systems simultaneously. Another promising avenue is video game-based rehabilitation. These have been utilized for motor, cognitive, and swallowing training. Similarly, tele-rehabilitation provides access to speech therapy or cognitive exercises, enabling patients to address language and memory deficits from their homes. Various mobile applications have been developed that help in the rehabilitation of visual, cognitive, and communicative deficits, which provide low-cost, accessible, and affordable options to patients. This chapter will help the readers to analyze the evidence supporting the efficacy of different rehabilitation technologies across motor, sensory, cognitive, vision, speech, swallowing, and sphincter domains.

9.1 Need for Technology and Low-Cost Solutions

Stroke is one of the leading causes of disability globally, with an increasing proportion of survivors needing long-term rehabilitation. Post-stroke disability often extends beyond motor impairments to encompass sensory deficits, cognitive dysfunction, visual disturbances, speech and language difficulties, swallowing impairments, sphincter deficits, and emotional challenges. These diverse impairments significantly impact a survivor's quality of life, making comprehensive rehabilitation essential for recovery. Traditional rehabilitation frequently relies on one-on-one therapy performed by skilled rehabilitation professionals, which is difficult to scale in areas with a dearth of therapists. This gap highlights the critical need for developments in technology that would make rehabilitation more accessible and efficient.

The use of technology in stroke rehabilitation is increasingly recognized as essential for enhancing recovery outcomes. It can address the multifaceted nature of post-stroke disabilities, but high costs often limit accessibility, especially in low- and middle-income countries. This creates a major barrier to equitable health care in low-resource settings. There is an urgent need for low-cost, scalable solutions that offer similar benefits without the financial burden. Locally manufactured robotic systems, non-immersive virtual reality (VR), contextualized video-game-based rehabilitation, cost-effective telemedicine, and open-source software can help bridge this gap. Continued innovation will be key to meeting the growing global demand for affordable, effective stroke rehabilitation.

9.2 Technologies for Sensorimotor Rehab

Sensorimotor functions are most often impacted by conditions such as stroke, which have a debilitating effect on a person's independence and quality of life. Work on developing and clinically validating technology for sensorimotor neurorehabilitation has been ongoing for almost the past three decades. These technologies include robotic devices, sensor-based systems, computer vision-based systems, EMG Biofeedback systems, and neuromuscular electrical stimulation for assistive and neuromodulation purposes. Each technology type caters to specific patient groups and addresses specific requirements for sensorimotor neurorehabilitation.

9.3 Robotics

Robots are best suited for relatively low-functioning patients who require assistance for training relatively simple movements. Robots have been developed and validated for neurorehabilitation of the upper limb, lower limb, and balance functions. Robots are well suited for the assisted training of relatively simple movements with game-based feedback for motivation and engagement. More affordable robotic solutions have recently started appearing, especially upper-limb rehabilitation robots, as there is no evidence supporting complex multi-degree-of-freedom robots, at least for the upper limb. These recent solutions are simpler, compact, more

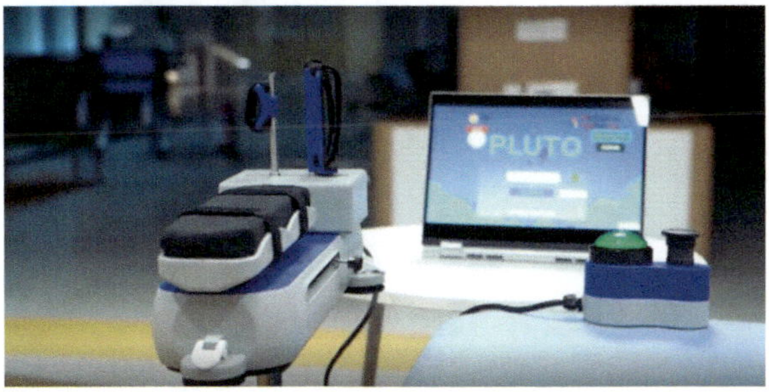

Fig. 9.1 PLUTO: low-cost upper limb robot

affordable, and could be used both in-clinic and at-home. Examples of these robots include the H-Man [1], PLUTO [2], PRIDE [3], RYMO, etc. These are robots specifically designed for training simple one- or two-degree-of-freedom movements of the arm (H-Man, MARS) or the wrist/hand (PLUTO (Fig. 9.1), PRIDE, RYMO) with physical assistance. These simple robots for focused training in various environments (hospitals, clinics, or patients' homes) are expected to increase therapy intensity significantly in the coming years. The G Gaiter from GenRobotics is a relatively affordable option for the lower limb. An indigenous wearable lower-limb robot exoskeleton is also being developed by the All India Institute of Medical Sciences (AIIMS) in partnership with DRDO and IIT Delhi to aid stroke and spinal cord injury patients in regaining their mobility.

Similarly, Mobility Assistive Gait Device for Rehabilitation (MITRA) (Fig. 9.2) has been designed and developed as an active exoskeleton for lower limbs, including actuated hip, knee, and ankle joints. The prime components of MITRA are: The Exoskeleton Suit, Control Hardware, Interfacing Circuit Board, Touch Panel, and Power System. This exoskeleton device can be used as a capability magnifier or as an assisting device for spinal cord injury patients, stroke patients, and the elderly.

Depending on the degree of disability, robotics can be used to give tailored interventions as well as to evaluate performance before, during, and after an intervention [4]. Insights from a systematic review and meta-analysis conducted in 2012 emphasize its efficacy and potential for transforming post-stroke care. Robot-assisted therapy (RAT) demonstrated significant improvements in upper limb motor function [5]. RAT showed a statistically significant improvement in outcomes like Fugl-Meyer assessment (FMA) and WMFT compared to conventional therapy, with an overall SMD of 0.46 (95% CI). Another recent systematic review (2020) showed a slight advantage of RAT over conventional therapy with SMD 0.40 [6].

In the case of the lower limb, robotic devices aid in gait rehabilitation, body weight supported training (BWSTT), and direct leg movements to resemble natural walking. BWSTT offers improvement in gait but has no significant advantage over conventional gait training strategies for chronic stroke survivors [7].

The review consolidates findings from randomized controlled trials and meta-analyses, highlighting improvements in parameters like gait speed, endurance, and muscle strength when using robotic systems with an overall SMD ranging from

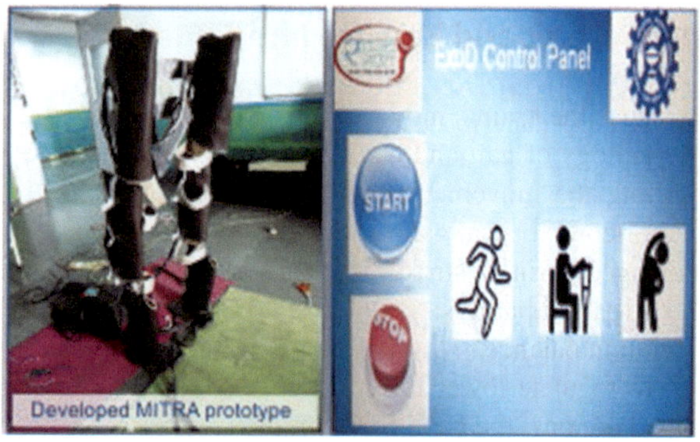

Fig. 9.2 MITRA prototype

0.30 to 0.80 and a CI of 95%. In another systematic review (2024), it was shown that the relationship between the effectiveness of robot-assisted training and time since stroke was statistically significant ($p = 0.001$). Robot-assisted training for the lower limb has a small-to-moderate positive effect on balance (SMD = 0.25), which is statistically significant given the 95% CI excludes zero [8].

9.4 Sensor-Based Systems

Patients with sufficient residual movements often don't require physical assistance. Sensor-based systems, designed to measure movement, interaction, and provide feedback, offer valuable support across upper limb, lower limb, and balance neurorehabilitation. Some examples of sensor-based systems for upper limb neurorehabilitation include Armeo Spring, Pablo, Diego, BeAble, Myro, gripAble [9], MusicGlove [10], Fit.Me [11], etc. (Fig. 9.3). Similar systems for balance neurorehabilitation also exist, which provide a standing surface that can measure and provide feedback on the center of pressure for training assessment purposes. Some popular examples of sensor-based systems for balance rehabilitation include the Nintendo Wii Board [8], the Balance Master, Hawkin Dynamics, Bertec Force Plates, etc. More recent wearable systems for movement tracking and feedback also fall in this category. In general, sensor-based systems are well suited for training more complex movements for higher-functioning patients.

Sensor-based systems for motor rehabilitation, particularly in stroke recovery, use a variety of technologies such as Inertial Measurement Units and motion capture systems to monitor and improve motor function [12]. In a study using the Gloreha Aria sensor-based system, the change in FMA-UE scores for the treatment group was 12.7 ± 8.6, and for the conventional group, it was 11.8 ± 9.2 after 30 rehabilitation sessions [13]. Dynamic posturography or force platform with visual feedback technique is an effective approach for post-stroke balance training, as it

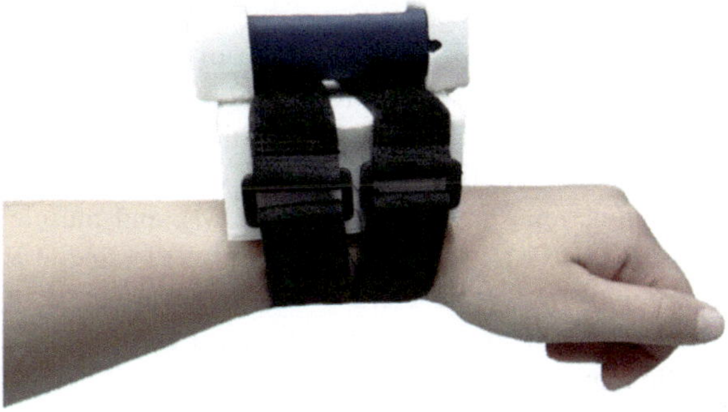

Fig. 9.3 T'ena sensor system for upper limb rehab

results in better locomotor abilities. This mode of training is well tolerated by patients with stroke and is compatible with rehabilitation practices in a clinical setting [14].

Therefore, sensor-based rehabilitation technologies have shown potential in improving motor function in stroke patients, though further studies with larger sample sizes and more refined outcome measures are needed to solidify the evidence.

9.5 Computer Vision-based Systems

Computer vision-based systems can be used for whole-body complex movements without requiring any additional instrumentation. These are mostly research prototypes, and a few commercial systems are currently available for neurorehabilitation applications. The Kinect-based system from MindMaze is a good example of a computer vision-based system for neurorehabilitation applications. These systems are ideal for quantifying movement kinematics but do not measure physical interaction with objects.

Computer vision-based systems have demonstrated encouraging results in stroke patients' motor rehabilitation, particularly upper extremity rehabilitation [15]. While some studies report improvements in motor functions, including hand positioning and coordination, the detailed statistical outcomes (such as SMD or CI values) are generally absent or need further standardization across trials.

9.6 Virtual Reality

VR has the potential to transform stroke rehabilitation in LMICs by offering accessible, cost-effective alternatives. It enables immersive, engaging therapy that improves motor recovery through gamified, repetitive exercises tailored to individual needs. VR helps overcome limited access to specialized therapists and facilities, allowing high-quality rehabilitation in resource-constrained settings. Portable systems support home-based therapy, reducing travel burdens for patients and caregivers. Evidence shows VR enhances motivation, adherence, and outcomes, helping bridge rehabilitation service gaps. Integrating VR can advance stroke care in LMICs, ensuring equitable recovery opportunities.

VR is a widely used technology in post-stroke motor rehabilitation. A 2017 Cochrane systematic review found VR more effective than no intervention for upper limb function, with small but significant improvements measured by the Fugl-Meyer assessment (FMA) [16]. Chen et al. reported significant gains in upper limb motor function, range of motion (ROM), and muscle strength using VR therapy, measured by FMA-UE and manual muscle testing [17]. Fang et al. found VR combined with traditional therapy was more effective than traditional rehabilitation alone for upper limb recovery, as measured by FMA-UE and manual dexterity on the Box and Block Test [18].

Two systematic reviews on VR in the subacute phase of stroke found no significant motor function improvements compared to conventional therapy [19, 20].

Further research is needed to assess its effectiveness in the acute and sub-acute phases.

9.7 Neuromuscular Electrical Stimulation

Neuromuscular electrical stimulation (NMES) systems are suited for assisted training of simple movements and neuromodulation applications [21]. Robotic technology is generally more complex, expensive, and less portable, while other options are simpler, compact, portable, and cheaper. These systems can assist functional movements (e.g., reach and grasp) through functional electrical stimulation (FES) or support simpler, nonfunctional movements for strength training and maintaining muscle bulk. Transcutaneous electrical stimulation (TENS) provides sensory stimulation for neuromodulation and pain relief. Newer techniques include transcranial direct current stimulation (TDCS), epidural spinal cord stimulation, and vagal nerve stimulation, explored for facilitating or relearning movements [22–25].

NMES is typically divided into two categories: FES and TENS [26, 27]. FES activates muscle fibers to produce contractions and movement, while TENS targets sensory nerve fibers and is commonly used for pain management. FES can be applied via pre-programmed patterns without feedback or combined with feedback-based systems involving active patient participation, such as EEG-FES or EMG-FES (Fig. 9.4).

Two systematic reviews report NMES (FES and TENS) benefits in improving lower limb motor function and balance. Hong et al. (2018) found significant gains in lower limb function, gait speed (10 MWT), and reduced TUG times. Chen et al.

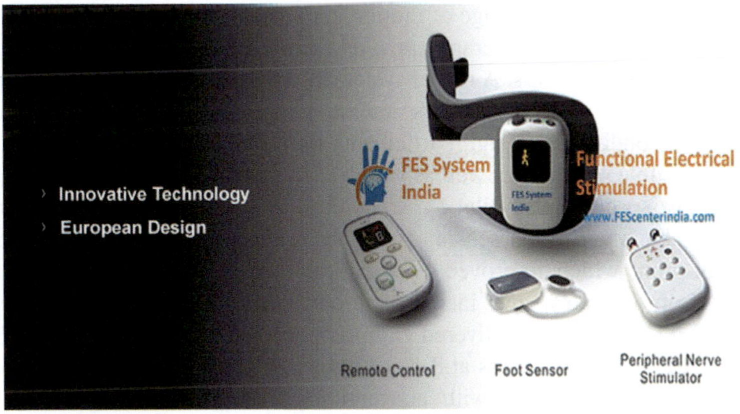

Fig. 9.4 FES systems

(2024) reported improvements in gait speed, cadence, step length, ankle dorsiflexion, and 6-min walk test distance, with greater effects in chronic patients and early interventions (within 4 weeks). NMES also improved Berg Balance Scale (BBS) and TUG scores compared to controls.

FES is effective for upper limb rehabilitation. Kim et al. (2023) combined FES with mirror therapy, achieving superior results in the Fugl-Meyer Assessment, Manual Function Test, and range of motion compared to mirror therapy alone ($p < 0.05$). A systematic review confirmed FES benefits for upper limbs, especially when combined with EMG-FES or EEG-FES systems [28–30].

9.8 Mobile Health

mHealth apps are transforming stroke care in LMICs by addressing limited healthcare access and resources. Examples include System-Integrated techNology-Enabled Model of cAre (SINEMA) in rural China, which trains village doctors and enhances stroke prevention through mobile technology [31]; Care for Stroke, which supports rehabilitation and self-management with guided exercises and educational tools [32]; and Phone-based Interventions under Nurse Guidance after Stroke (PINGS) in Ghana, which enables nurses to monitor and guide blood pressure management post-stroke [33]. Other mHealth rehabilitation solutions aim to deliver cost-effective rehabilitation via mobile platforms, overcoming infrastructure barriers. While promising for underserved populations, challenges remain, including smartphone penetration, internet connectivity, and cultural adaptability. Integrating mHealth into healthcare systems can help improve outcomes and bridge the stroke care gap in LMICs.

In stroke rehabilitation, mHealth can improve motor function in gait, balance, upper and lower extremity range of motion, and hand dexterity. Most interventions use one of four app types: therapy apps, rehabilitation videos, reminders, or combined videos and reminders. A recent systematic review found significant improvements in temporal gait parameters and dynamic balance in individuals with neurological conditions, though participants included those with Parkinson's disease and multiple sclerosis as well as stroke [34].

For stroke-specific rehabilitation, Chae et al. developed a home-based system using wearable smartwatch sensors for exercise monitoring and therapist-patient communication, which improved Wolf Motor Function Test scores ($p = 0.02$), reduced drop-out rates, and increased motivation [35]. An 8-week caregiver-mediated exercise program delivered via app showed no significant change in Stroke Impact Scale—mobility scores at 8 or 12 weeks [36]. Salgueiro et al. (2022) combined app-guided exercises with conventional therapy, achieving a 2.76-point improvement in the Trunk Impairment Scale compared to controls [37].

9.9 Technologies for Visual Rehab

Visual impairments following stroke may include abnormalities of central and/or peripheral vision, eye movements, and a variety of visual perception problems, such as inattention and agnosia [38]. Affordable technologies effectively support vision rehabilitation post-stroke by enhancing visual function and adapting to field deficits. Tablet and smartphone apps provide structured home exercises for scanning and attention, while prism glasses and lens adaptations passively address issues like hemianopia, though adaptation time may be required [39]. Low-cost VR devices, such as Google Cardboard, offer immersive eye movement training, and homemade visual scanning boards encourage regular scanning with simple materials. Certain programs like SightScience and ReadRight deliver customizable, often free, computer-based exercises. Printable activities, such as word searches and object-tracking tasks, build spatial awareness, while practical adjustments like improved lighting and contrast improve safety [40]. Combining these tools with traditional therapy enables a comprehensive, accessible approach to vision recovery (Fig. 9.5).

Vision typoscopes, Peli prisms, reading aids, and scanning therapy for stroke patients with field loss, with advice on head postures, are the most frequently reported strategies [41]. Peli prism lens system offers a measurable field expansion of up to 20° at primary gaze without causing double vision in central vision. It is relatively inexpensive and cosmetically acceptable, and double vision is limited to the periphery [42].

Google Cardboard is a low-cost device that uses a smartphone to create virtual reality. Saraiva et al. used an interactive virtual environment for rehabilitating patients with mechanical strabismus and amblyopia. Developed with free software and low-cost virtual reality glasses, including Google Cardboard, the method proved to be a feasible and accessible treatment for these conditions [39]. Free smartphone apps with magnification tools as digital aids provide improvement in VA and reading performance. Also, free smartphone apps for magnification should be considered in reading rehabilitation programs for low vision patients [43].

Fig. 9.5 Vision aids: Peli prism, spectacle

Systematic audio–visual stimulation might induce a long-lasting amelioration of visual field disorders with patients showing improved visual detection and visual oculomotor exploration following training [40].

9.10 Technologies for Cognitive Rehabilitation

Various affordable tools support cognitive rehabilitation post-stroke. Mobile apps like Lumosity, Peak, and CogniFit provide exercises for memory, attention, and problem-solving, with free or low-cost options. Low-cost VR devices, such as Google Cardboard, offer immersive cognitive training for spatial awareness and attention. Tablets can run programs like RehaCom or BrainHQ, designed for rehabilitation, and facilitate video modeling for task sequencing and recall [44]. Wearable technology, including smartwatches and fitness trackers, aids routine management, reminders, and physical activity tracking, thereby supporting executive function and sleep quality [45]. Simple aids like whiteboards, sticky notes, audio recorders, and puzzle books (e.g., Sudoku, crosswords) enhance memory and problem-solving (Fig. 9.6). Community platforms, such as Facebook groups, provide social support, while telehealth services like Doxy.me and Zoom enable remote therapy.

Virtual reality (VR) and computer-assisted cognitive rehabilitation (CACR) are two emerging rehabilitation therapies for stroke patients. CACR is a computer-based cognitive training program with exercises and games using multimedia resources to enhance memory, attention, problem-solving, work simulation, language, practice, and processing speed. VR immerses users in a computer-generated world that responds to head and body movements through computers, head-mounted displays, tracking sensors, and real-time graphics [46].

Systematic review by Ene et al. provided preliminary evidence that technology-based cognitive rehabilitation interventions are feasible and acceptable for research in a stroke population [44]. The vast majority of mobile health apps are not

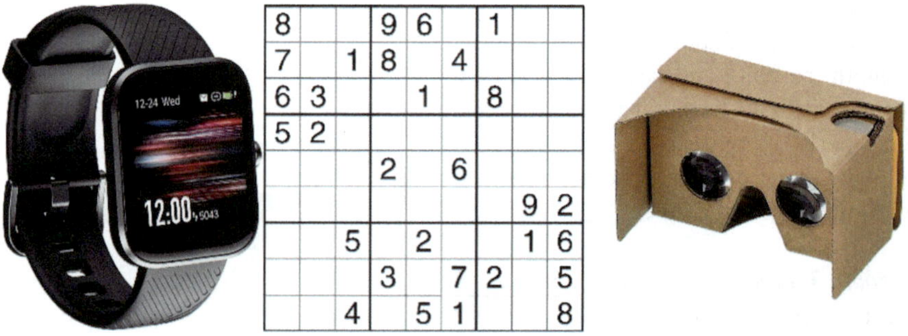

Fig. 9.6 Cognitive rehab aids—smart watch, crosswords, Google Cardboards, etc.

evidence-based at present [45]; however, a small but growing literature is now speaking to the use of smart technologies to both help improve functioning and monitor functioning for persons with acquired brain injury. Mingming et al. conducted a systematic review on computer-based training for post-stroke cognitive rehabilitation, finding it not significantly superior to traditional methods [47].

9.11 Technologies for Speech Rehab

Affordable mobile apps like Tactus Therapy, Lingraphica, and Constant Therapy provide structured speech exercises for various language skills, while communication apps like Voice4u and LetMeTalk support expression through picture symbols [48]. Tablets are versatile therapy hubs, compatible with many speech apps and helpful for interactive sessions with family [49]. Voice assistants such as Alexa or Siri enable speech practice through simple commands, reinforcing language skills in a relaxed setting. Simple augmentative communication tools, like handmade communication boards or affordable AAC apps, also help those with severe language challenges to express themselves. Websites like Aphasia Toolbox offer free printable games that encourage repetition in an enjoyable way. To combat the psychosocial effects of aphasia, including social isolation, mobile devices that help people connect to others like FaceTime, mail can also be used [34]. Jellow (Fig. 9.7) is a free app developed at IIT Bombay that helps people with speech difficulties, available in multiple languages. Other applications include Speech Assistant AAC, Avaz, Commboards, etc.

A systematic review of seven studies on computer therapy for aphasia found it effective for improving language skills, including comprehension, sentence production, word-finding, and reading, compared to no therapy [50]. Popular gaming apps can effectively support executive functions. The Speech Therapy App (STAPP), a language exercise program for aphasia, was tested on 25 acute post-stroke participants and proved feasible as an additional rehabilitation tool alongside standard care [51]. Computer-based training sessions resulted in more significant speech

Fig. 9.7 Communication applications—Jellow

improvements compared to traditional treatments [48]. Mobile application-based therapy is gaining attention as a means of remote delivery [52]. The advantage of this treatment modality is its potential for increased cost-effectiveness, reduced therapist burden, and enhanced patient satisfaction and treatment adherence [53]. RecoverNow, a mobile tablet-based post-stroke communication therapy, was studied in acute patients, and they showed interest in the initial stages too [49].

9.12 Technologies for Swallow Rehab

Affordable options like surface EMG (sEMG) biofeedback devices, such as MyoWare muscle sensors, provide patients with real-time visual feedback on muscle activity, enabling home-based exercises [54]. Mobile apps like Swallow Prompt and SwallowIT deliver guided swallowing exercises and tracking, while teletherapy platforms allow remote therapy sessions. Low-cost pressure sensors, including DIY Arduino-based devices, help patients visualize their exercise accuracy. Gamified therapy, using low-cost VR headsets and smartphone-based games, increases patient engagement and motivation. Simple resistance bands or tongue depressors can replace expensive tools like TheraBite for muscle-strengthening exercises. Portable TENS units can offer affordable alternatives to neuromuscular electrical stimulation (NMES) therapy, but should be used under professional supervision (Fig. 9.8) [55].

Swallow effects of home-based dysphagia rehabilitation using Information Communication Technology (ICT) on tongue muscle strength and volume showed effects similar to those of conventional swallowing rehabilitation. Therefore, it can be conducted at home without the help of a therapist [56]. Community-based rehabilitation program is more effective than traditional treatment in improving quality of life and depressive symptoms, but have similar

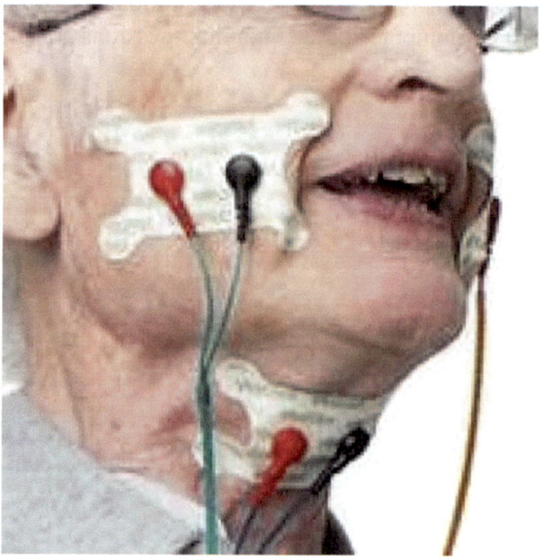

Fig. 9.8 TENS for dysphagia rehabilitation

efficacy to that of conventional rehabilitation in swallowing function [57]. EMG biofeedback as an adjunctive therapy exerts positive effects on swallow function [54]. Video game-based swallowing function training systems were generally accepted, and patients were willing to use the system for dysphagia rehabilitation [58]. Chung et al. developed an integrated rehabilitation APP that allows patients to follow the video demonstrations and perform rehabilitation activities anytime and anywhere [59].

9.13 Technologies for Bladder/Bowel Rehab

Low-cost technologies effectively manage neurogenic bladder symptoms after a stroke, improving quality of life. Biofeedback devices help control pelvic floor muscles [60], while timed voiding and bladder training apps aid in establishing schedules, tracking patterns, and fluid intake. Simple tools like mirrors and TENS units enhance muscle strength and management (Fig. 9.9) [61]. Wearable sensors reduce incontinence by alerting users when the bladder is full, while apps encourage regular pelvic floor exercises.

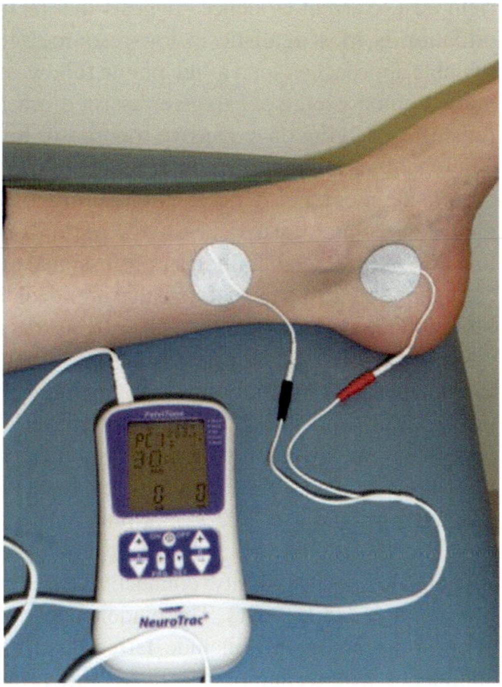

Fig. 9.9 TENS unit for transcutaneous tibial nerve stimulation

Similarly, low-cost tools support neurogenic bowel rehabilitation, promoting routine and independence. Timed reminders encourage regular bowel movements, while abdominal massage tools and TENS units enhance motility. Affordable enema kits, hydration tools, and portable bidets improve hygiene. Exercise reminders boost activity for better motility, even with limited mobility. TAI offers an effective, low-cost defecation method [62].

Remote long-term follow-up and management of neurogenic bladder and bowel dysfunction have been established by specialized health services using many different types of telecommunication technology [63]. App-based audio guidance pelvic floor muscle training on the treatment of stress urinary incontinence was found to be more effective and easier to comply with than home-based pelvic floor muscle training [64]. Transcutaneous tibial nerve stimulation [65], posterior tibial nerve stimulation [66] has emerged as a conservative, alternative treatment option for detrusor overactivity [67]. Neuromodulation by TENS reduces sensory input and inhibits motor reflex circuits in the pontine micturition center at a relatively low amplitude and modulates the neural activity of several pathways involved in bladder control [61].

Guide for Rehab Practitioners

There are new devices every day and the challenge is finding the right device/system for the right person in the right location at the right time in their recovery at the right cost. Current evidence suggests that technology-enhanced post-stroke rehabilitation is most realistic in low-resource settings when it builds from simple, scalable approaches such as telephone follow-up, basic smartphone support, video-guided home exercise, caregiver-assisted practice, and asynchronous messaging, rather than relying on expensive robotics or highly specialized platforms [68–70]. Telerehabilitation can improve access to rehabilitation after stroke, but it should be used to augment, not replace, essential in-person assessment and therapy, with the choice of technology matched to the person's goals, impairments, safety, and context [68, 70]. For rehabilitation professionals and patients trying to navigate this confusing space, the most defensible approach is to start with the local stroke rehabilitation team or service, then align decisions with national stroke guidelines and international implementation frameworks such as the World Stroke Organization Roadmap, rather than relying on commercial app stores or isolated product claims [68, 71, 72]. This is especially important in LMICs, where workforce shortages, uneven service availability, financing gaps, and limited community rehabilitation infrastructure mean that technologies must fit within existing systems and be adapted for local use [70, 72]. A pragmatic clinical screen should include functional goals, digital literacy, language, cognition, communication ability, caregiver support, device availability, connectivity, and affordability before any technology is recommended [69, 71]. The evidence also shows that barriers to mHealth use after stroke commonly include fatigue and other health-related limitations, low confidence with technology, infrastructure problems, inadequate support, and time constraints, so usability and support matter as much as efficacy [71]. What the field still lacks is a trusted central repository for stroke mHealth that combines evidence quality with a pragmatic implementation lens and consumer-facing information on usability, cost, and perceived value [71, 72]. NeuroRehabDirectory.com is a useful

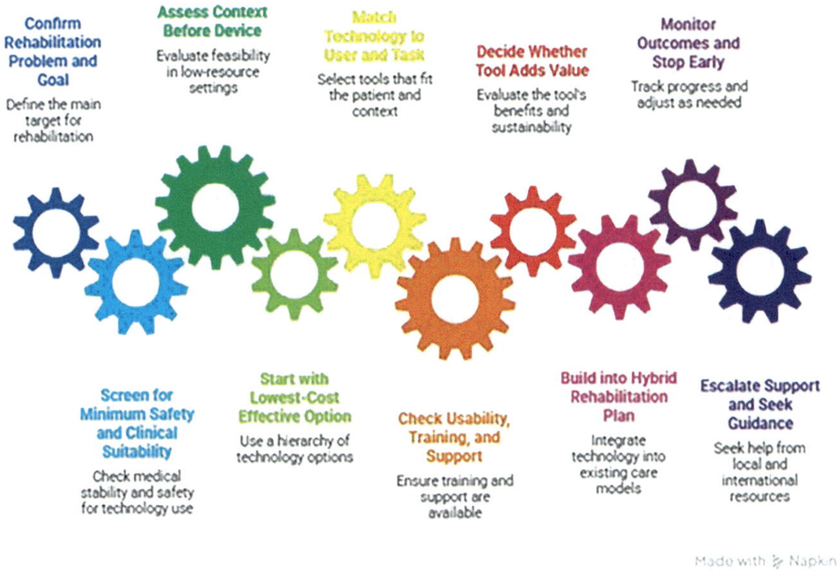

Fig. 9.10 Step decision making tool

example of the type of searchable, consumer-oriented directory that begins to fill this gap, but there remains a clear need for a broader, independently curated, LMIC-relevant repository to guide both clinicians and patients [73].

10-Step Decision Making Tool (Fig. 9.10)

Step 1: Confirm the rehabilitation problem and goal. Define the main target first: mobility, upper limb use, communication, cognition, swallowing, self-care, participation, caregiver support, or exercise adherence. Technology should only be considered if it solves a clearly identified rehabilitation problem and aligns with patient goals.

Step 2: Screen for minimum safety and clinical suitability. Before recommending any technology, check medical stability, fall risk, cognition, communication, vision, hearing, neglect, fatigue, mood, seizure risk where relevant, and whether remote or partially remote rehabilitation is safe for that individual.

Step 3: Assess the context before the device. In low-resource settings, the key question is not "What technology is available?" but "What is feasible here?" Assess smartphone access, electricity, internet reliability, language, digital literacy, caregiver availability, transport burden, cost to family, and local follow-up capacity.

Step 4: Start with the lowest-cost effective option. Use a "least complex first" hierarchy: (a) phone call/SMS/WhatsApp support, (b) prerecorded or live video-guided home exercise, (c) caregiver-supported structured practice, (d) simple wearable or app-based monitoring, and only then (e) advanced platforms or devices if clearly justified and sustainable.

Step 5: Match the technology to the user and the task. Select only tools that fit the person's impairment profile, goals, literacy, home environment, and support system. The patient should not be forced to adapt to the technology; the technology should adapt to the patient and context.

Step 6: Check usability, training, and support. Do not implement any tool unless the patient, caregiver, and rehabilitation team can be trained to use it. Evidence and guidelines emphasize that both stroke survivors and staff need support in using telerehabilitation technologies safely and appropriately.

Step 7: Decide whether the tool adds value beyond usual care. Ask four questions: Does it improve access? Does it increase dose/intensity or adherence? Is it acceptable to the patient/caregiver? Is it affordable and sustainable locally? If the answer is "no" to most of these, do not proceed.

Step 8: Build it into a hybrid rehabilitation plan. Technology should usually be an adjunct to therapist-guided care, not a replacement for assessment, progression, troubleshooting, and safety review. Hybrid models are especially relevant in LMICs and other low-resource settings.

Step 9: Monitor outcomes and stop early if it is not working. Review function, adherence, safety events, user burden, caregiver burden, and dropout risk. If the tool is ineffective, too complex, too expensive, or poorly accepted, step down to a simpler option.

Step 10: Escalate support and seek trusted guidance. If clinicians or patients are unsure, they should first contact the local stroke rehabilitation team or service, then national stroke/rehabilitation societies, and then international implementation resources such as the World Stroke Organization. The field still lacks a trusted central repository linking evidence, implementation practicality, and consumer-facing value; this is a major unmet need, especially for LMICs. In low-resource settings, the best technology-enhanced post-stroke rehabilitation strategy is to start with the simplest scalable option that is safe, goal-directed, affordable, and supported by caregivers and clinicians, while using technology to extend rehabilitation rather than replace it.

9.14 Summary

This chapter highlights the critical role of technology-enhanced rehabilitation in addressing the multifaceted challenges of post-stroke recovery, particularly in low-resource settings. It addresses how affordable and scalable solutions can bridge gaps in access to rehabilitation services while covering interventions for sensorimotor, cognitive, visual, speech, and sphincter dysfunctions (Table 9.1).

Technologies like robotics, wearable sensors, and virtual reality (VR) significantly enhance motor recovery (Table 9.2). Neuromuscular electrical stimulation (NMES), including functional electrical stimulation (FES), aids in both motor function recovery and neuromodulation, offering versatile applications in stroke care.

Cognitive impairments are addressed using apps like Lumosity and BrainHQ, low-cost VR like Google Cardboard, and tools such as smartwatches, which enhance

Table 9.1 Interventions for stroke rehabilitation

Domain	Interventions
Sensorimotor	Robotics Sensor-based systems Virtual reality NMES Mobile health
Vision	Typoscopes Peli prisms Reading aids Google cardboards
Cognition	Virtual reality Computer-assisted cognitive rehabilitation Mobile apps Wearable technology Community platforms
Communication	Mobile health Tablet-based applications Augmentive communication tools
Swallow	EMG biofeedback TENS Virtual reality Mobile applications Community-based platforms
Sphincter	TENS Biofeedback Electrical stimulation Mobile apps Diaries, Oenema kits, hydration tools, etc

focus, memory, and independence. Stroke survivors with visual impairments benefit from prism glasses, scanning therapy, VR, and apps like SightScience, improving visual awareness and coordination. Speech rehabilitation utilizes apps like Tactus Therapy, voice assistants, and communication aids to rebuild language skills and social connections. Dysphagia and incontinence are managed with tools like SwallowIT, TENS units, biofeedback devices, and wearable sensors, improving safety, coordination, and independence.

In conclusion, this chapter demonstrates that technology-enhanced rehabilitation offers effective solutions for addressing post-stroke impairments across multiple domains. By prioritizing affordability and accessibility, these innovations can significantly improve the quality of life for stroke survivors, especially in underserved regions. The integration of these tools into healthcare systems underscores the potential for equitable, scalable, and impactful rehabilitation practices worldwide.

Table 9.2 What does evidence say (WSO guidelines 2025)

What does evidence say:

- In people with post-stroke arm weakness, strengthening and motor recovery may be supported through repetitive task practice delivered with assistive devices, constraint-induced movement therapy, and robotic training
- Electrical stimulation can be considered for stroke survivors with weakness in the upper or lower limbs, especially when muscle power is less than antigravity strength
- For individuals with mild to severe weakness of the upper limb, mechanically assisted arm training, including robotic devices, may be used to improve arm and hand function
- Virtual reality-based rehabilitation and interactive gaming approaches may be incorporated to support recovery of upper limb function
- In patients with mild to severe weakness of the arm or hand, electrical stimulation used alongside active motor retraining may help improve upper limb performance
- Non-invasive brain stimulation techniques, such as transcranial direct current stimulation and repetitive transcranial magnetic stimulation, are not recommended as part of routine care for improving arm function and should be confined to research use
- These brain stimulation techniques are also not recommended in usual clinical practice for improving activities of daily living and should only be applied within research settings
- In chronic post-stroke aphasia, transcranial direct current stimulation or repetitive transcranial magnetic stimulation, whether combined with conventional aphasia therapy or not, should not be used routinely for speech and language rehabilitation outside research
- For stroke survivors at risk of shoulder subluxation, electrical stimulation may be considered within the first 6 months after stroke to help prevent or reduce subluxation
- For those experiencing post-stroke shoulder pain, electrical stimulation may be used as part of pain management
- In stroke survivors who have difficulty standing, balance-focused task practice may be supplemented with other approaches such as:
    1. Virtual reality-based training
    2. Treadmill-based balance or gait work
    3. Motioncapture or force-sensing systems, including devices such as balance boards
    4. Visual or auditory feedback, including force-platform biofeedback
    5. Electromechanically assisted standing or gait training
- Individuals with walking impairment after stroke should be offered as much structured, repetitive, task-specific walking practice as feasible, including practice of components of walking where needed
- Methods that may be used to support gait retraining include:
    1. Circuit class therapy, particularly with emphasis on overground walking
    2. Treadmill training, with or without bodyweight support
- In addition to standard walking practice, other options that may be considered for gait rehabilitation include:
    1. Virtual reality training
    2. Electromechanically assisted gait training
    3. Biofeedback
    4. Cadence cueing
    5. Electrical stimulation
- Telehealth may be used as an alternative means of delivering rehabilitation, particularly for patients who cannot easily access specialist community rehabilitation services
- Telehealth may also be used alongside in-person rehabilitation, provided the specific intervention being delivered remotely has evidence to support its benefit

References

1. Hussain A, Dailey W, Hughes C, Tommasino P, Budhota A, Gamage WGKC, et al. Preliminary feasibility study of the H-Man planar robot for quantitative motor assessment. In: 2015 IEEE/RSJ international conference on intelligent robots and systems (IROS). New York: IEEE; 2015. p. 6167–72. Available from: https://ieeexplore.ieee.org/abstract/document/7354256.
2. Nehrujee A, Andrew H, Reethajanetsurekha PA, Samuelkamaleshkumar S, Prakash H, et al. Plug-and-train robot (PLUTO) for hand rehabilitation: design and preliminary evaluation. IEEE Access. 2021;9:134957–71.
3. Rätz R, Conti F, Müri RM, Marchal-Crespo L. A novel clinical-driven Design for Robotic Hand Rehabilitation: combining sensory training, effortless setup, and large range of motion in a palmar device. Front Neurorobot. 2021:15. Available from: https://www.frontiersin.org/journals/neurorobotics/articles/10.3389/fnbot.2021.748196/full.
4. The Present and Future of Robotic Technology in Rehabilitation | Current Physical Medicine and Rehabilitation Reports [Internet]. [cited 2024 Dec 26]. Available from: https://link.springer.com/article/10.1007/s40141-016-0139-0.
5. Norouzi-Gheidari N, Archambault PS, Fung J. Effects of robot-assisted therapy on stroke rehabilitation in upper limbs: systematic review and meta-analysis of the literature. Washington: U.S. Department of Veteran's Affairs.
6. Robot-assisted therapy for upper-limb rehabilitation in subacute stroke patients: a systematic review and meta-analysis - Chien - 2020 - Brain and Behavior - Wiley Online Library [Internet]. [cited 2024 Dec 26]. Available from: https://doi.org/10.1002/brb3.1742.
7. Srivastava A, Taly AB, Gupta A, Kumar S, Murali T. Bodyweight-supported treadmill training for retraining gait among chronic stroke survivors: a randomized controlled study. Ann Phys Rehabil Med. 2016;59(4):235–41.
8. Effectiveness of robot-assisted lower limb rehabilitation on balance in people with stroke: a systematic review, meta-analysis, and meta-regression | SpringerLink [Internet]. [cited 2024 Dec 26]. Available from: https://link.springer.com/chapter/10.1007/978-3-031-59091-7_7.
9. Mace M, Mutalib SA, Ogrinc M, Goldsmith N, Burdet E. GripAble: an accurate, sensitive and robust digital device for measuring grip strength. J Rehabil Assist Technol Eng. 2022;9:20556683221078455.
10. Friedman N, Chan V, Reinkensmeyer AN, Beroukhim A, Zambrano GJ, Bachman M, et al. Retraining and assessing hand movement after stroke using the MusicGlove: comparison with conventional hand therapy and isometric grip training. J NeuroEng Rehabil. 2014;11(1):76.
11. Ramos Muñoz EDJ, Swanson VA, Johnson C, Anderson RK, Rabinowitz AR, Zondervan DK, et al. Using large-scale sensor data to test factors predictive of perseverance in home movement rehabilitation: optimal challenge and steady engagement. Front Neurol. 2022;13. Available from: https://www.frontiersin.org/journals/neurology/articles/10.3389/fneur.2022.896298/full.
12. Nascimento LM, Bonfati LV, Freitas MLB, Mendes Junior JJA, Siqueira HV, Stevan SL. Sensors and systems for physical rehabilitation and health monitoring—a review. Sensors. 2020;20(15):4063.
13. Vanoglio F, Comini L, Gaiani M, Bonometti GP, Luisa A, Bernocchi P. A sensor-based upper limb treatment in hemiplegic patients: results from a randomized pilot study. Sensors. 2024;24(8):2574.
14. Srivastava A, Taly AB, Gupta A, Kumar S, Murali T. Post-stroke balance training: role of force platform with visual feedback technique. J Neurol Sci. 2009;287(1):89–93.
15. Sucar LE, Luis R, Leder R, Hernández J, Sánchez I. Gesture therapy: a vision-based system for upper extremity stroke rehabilitation. In: 2010 Annual International Conference of the IEEE Engineering in Medicine and Biology [Internet]. 2010 [cited 2024 Dec 26]. p. 3690–3. Available from: https://ieeexplore.ieee.org/abstract/document/5627458.

16. Laver K. Virtual reality for stroke rehabilitation. In: Virtual reality in health and rehabilitation. Boca Raton: CRC Press; 2020.
17. Chen J, Or CK, Chen T. Effectiveness of using virtual reality–supported exercise therapy for upper extremity motor rehabilitation in patients with stroke: systematic review and meta-analysis of randomized controlled trials. J Med Internet Res. 2022;24(6):e24111.
18. Fang Z, Wu T, Lv M, Chen M, Zeng Z, Qian J, et al. Effect of traditional plus virtual reality rehabilitation on prognosis of stroke survivors: a systematic review and meta-analysis of randomized controlled trials. Am J Phys Med Rehabil. 2022;101(3):217.
19. Zhang B, Wong KP, Kang R, Fu S, Qin J, Xiao Q. Efficacy of robot-assisted and virtual reality interventions on balance, gait, and daily function in patients with stroke: a systematic review and network meta-analysis. Arch Phys Med Rehabil. 2023;104(10):1711–9.
20. Hao J, Yao Z, Harp K, Gwon Y, Chen Z, Siu KC. Effects of virtual reality in the early-stage stroke rehabilitation: a systematic review and meta-analysis of randomized controlled trials. Physiother Theory Pract. 2023;39(12):2569–88.
21. Neuromuscular electrical stimulation in neurorehabilitation - Sheffler - 2007 - Muscle & Nerve - Wiley Online Library [Internet]. [cited 2024 Dec 26]. Available from: https://onlinelibrary.wiley.com/doi/abs/10.1002/mus.20758.
22. Neuromuscular Stimulation for Upper Extremity Motor and Functional Recovery in Acute Hemiplegia | Stroke [Internet]. [cited 2024 Dec 26]. Available from: https://www.ahajournals.org/doi/full/10.1161/01.STR.29.5.975.
23. Multi-Field Surface Electrode for Selective Electrical Stimulation - Popović-Bijelić - 2005 - Artificial Organs - Wiley Online Library [Internet]. [cited 2024 Dec 26]. Available from: https://onlinelibrary.wiley.com/doi/abs/10.1111/j.1525-1594.2005.29075.x.
24. Popovic DB, Popovic MB, Sinkjær T. Neurorehabilitation of upper extremities in humans with sensory-motor impairment. Neuromodulation Technol Neural Interface. 2002;5(1):54–67.
25. Popovic DB, Popovic MB, Sinkjær T, Stefanovic A, Schwirtlich L. Therapy of paretic arm in hemiplegic subjects augmented with a neural prosthesis: a cross-over study. Can J Physiol Pharmacol. 2004 Jul;82(8–9):749–56.
26. Restitution of reaching and grasping promoted by functional electrical therapy - Popovic - 2002 - Artificial Organs - Wiley Online Library [Internet]. [cited 2024 Dec 26]. https://doi.org/10.1046/j.1525-1594.2002.06950.x.
27. Sonde L, Kalimo H, Fernaeus SE, Viitanen M. Low TENS treatment on post-stroke paretic arm: a three-year follow-up. Clin Rehabil. 2000;14(1):14–9.
28. Reis J, Robertson E, Krakauer JW, Rothwell J, Marshall L, Gerloff C, et al. Consensus: "can tDCS and TMS enhance motor learning and memory formation?". Brain Stimulat. 2008;1(4):363–9.
29. Epidural stimulation of the cervical spinal cord improves voluntary motor control in post-stroke upper limb paresis | medRxiv [Internet]. [cited 2024 Dec 26]. Available from: https://www.medrxiv.org/content/10.1101/2022.04.11.22273635v1.
30. Ward N. The prospects for poststroke neural repair with vagal nerve stimulation. J Neurol Neurosurg Psychiatry. 2023;94(4):255–6.
31. Gong E, Gu W, Sun C, Turner EL, Zhou Y, Li Z, et al. System-integrated technology-enabled model of care to improve the health of stroke patients in rural China: protocol for SINEMA—a cluster-randomized controlled trial. Am Heart J. 2019;207:27–39.
32. Sureshkumar K, Murthy GVS, Munuswamy S, Goenka S, Kuper H. 'Care for Stroke', a web-based, smartphone-enabled educational intervention for management of physical disabilities following stroke: feasibility in the Indian context. BMJ Innov. 2015;1(3). Available from: https://innovations.bmj.com/content/1/3/127.
33. Sarfo FS, Akpalu A, Bockarie A, Appiah L, Nguah SB, Ayisi-Boateng NK, et al. Phone-based intervention under nurse guidance after stroke (PINGS II) study: protocol for a phase III randomized clinical trial. J Stroke Cerebrovasc Dis. 2021;30(8):105888.
34. Szeto SG, Wan H, Alavinia M, Dukelow S, MacNeill H. Effect of mobile application types on stroke rehabilitation: a systematic review. J NeuroEng Rehabil. 2023;20(1):12.

35. Moreno-Ligero M, Lucena-Anton D, Salazar A, Failde I, Moral-Munoz JA. mHealth impact on gait and dynamic balance outcomes in neurorehabilitation: systematic review and meta-analysis. J Med Syst. 2023;47(1):75.
36. JMIR mHealth and uHealth - Development and clinical evaluation of a web-based upper limb home rehabilitation system using a smartwatch and machine learning model for chronic stroke survivors: prospective comparative study [Internet]. [cited 2024 Dec 27]. Available from: https://mhealth.jmir.org/2020/7/e17216/.
37. Vloothuis JDM, Mulder M, Nijland RHM, Goedhart QS, Konijnenbelt M, Mulder H, et al. Caregiver-mediated exercises with e-health support for early supported discharge after STROKE (CARE4STROKE): a randomized controlled trial. PLoS One. 2019;14(4):e0214241.
38. Kim YS, Song JY, Park SH, Lee MM. Effect of functional electrical stimulation-based mirror therapy using gesture recognition biofeedback on upper extremity function in patients with chronic stroke: a randomized controlled trial. Medicine. 2023;102(52):e36546.
39. Randomized Crossover Clinical Trial of Real and Sham Peripheral Prism Glasses for Hemianopia | Optics and Refraction | JAMA Ophthalmology | JAMA Network [Internet]. [cited 2024 Nov 1]. Available from: https://jamanetwork.com/journals/jamaophthalmology/fullarticle/1763096.
40. da Silva PBE, Leal AS, Ferraz NN. Usability of smartphone apps as reading aids for low vision patients. Disabil Rehabil Assist Technol. 2022;17(7):848–52.
41. Rowe F, Brand D, Jackson CA, Price A, Walker L, Harrison S, et al. Visual impairment following stroke: do stroke patients require vision assessment? Age Ageing. 2009;38(2):188–93.
42. Pollock A, Hazelton C, Brady M. Visual problems after stroke: a survey of current practice by occupational therapists working in UK stroke inpatient settings. Top Stroke Rehabil. 2011;18(Suppl 1):643–51.
43. Saraiva AA, Barros MP, Nogueira AT, Fonseca Ferreira NM, Valente A. Virtual interactive environment for low-cost treatment of mechanical strabismus and amblyopia. Information. 2018;9(7):175.
44. Xiao Z, Wang Z, Song G, Zhong Y, Zhang W. Rehabilitation efficacy comparison of virtual reality technology and computer-assisted cognitive rehabilitation in patients with post-stroke cognitive impairment: a network meta-analysis. J Clin Neurosci. 2022;103:85–91.
45. Ene CG. Investigating the feasibility and acceptability of technology-based cognitive rehabilitation after stroke [internet] [doctoral]. University of East Anglia. Norwich Medical School; 2023 [cited 2024 Nov 1]. Available from: https://ueaeprints.uea.ac.uk/id/eprint/93586/.
46. Aimola L, Lane AR, Smith DT, Kerkhoff G, Ford GA, Schenk T. Efficacy and feasibility of home-based training for individuals with homonymous visual field defects. Neurorehabil Neural Repair. 2014;28(3):207–18.
47. Maggio MG, Luca A, Calabrò RS, Drago F, Nicoletti A. Can mobile health apps with smartphones and tablets be the new frontier of cognitive rehabilitation in older individuals? A narrative review of a growing field. Neurol Sci. 2024;45(1):37–45.
48. De Cock E, Batens K, Feiken J, Hemelsoet D, Oostra K, De Herdt V. The feasibility, usability and acceptability of a tablet-based aphasia therapy in the acute phase following stroke. J Commun Disord. 2021;89:106070.
49. Mallet K, Shamloul R, Pugliese M, Power E, Corbett D, Hatcher S, et al. RecoverNow: a patient perspective on the delivery of mobile tablet-based stroke rehabilitation in the acute care setting. Int J Stroke. 2019;14(2):174–9.
50. Mingming Y, Bolun Z, Zhijian L, Yingli W, Lanshu Z. Effectiveness of computer-based training on post-stroke cognitive rehabilitation: a systematic review and meta-analysis. Neuropsychol Rehabil. 2022;32(3):481–97.
51. Zheng C, Lynch L, Taylor N. Effect of computer therapy in aphasia: a systematic review. Aphasiology. 2016;30(2-3):211–44.

52. Zhou Q, Lu X, Zhang Y, Sun Z, Li J, Zhu Z. Telerehabilitation combined speech-language and cognitive training effectively promoted recovery in aphasia patients. Front Psychol. 2018:9. Available from: https://www.frontiersin.org/journals/psychology/articles/10.3389/fpsyg.2018.02312/full.
53. Gerber SM, Schütz N, Uslu AS, Schmidt N, Röthlisberger C, Wyss P, et al. Therapist-guided tablet-based telerehabilitation for patients with aphasia: proof-of-concept and usability study. JMIR Rehabil Assist Technol. 2019;6(1):e13163.
54. Albuquerque LCA, Pernambuco L, da Silva CM, Chateaubriand MM, da Silva HJ. Effects of electromyographic biofeedback as an adjunctive therapy in the treatment of swallowing disorders: a systematic review of the literature. Eur Arch Otorrinolaringol. 2019;276(4):927–38.
55. Kushner DS, Johnson-Greene D, Cordero MK, Thomashaw SA, Rodriguez J. Swallowing outcomes and discharge destinations in acute stroke tube-feeding dependent dysphagia patients treated with neuromuscular electrical stimulation during inpatient rehabilitation. Am J Phys Med Rehabil. 2020;99(6):487.
56. Kim H, Kown Y, Choi H. Home based dysphagia rehabilitation for stroke patients using information and communication technology. J Int Acad Phys Ther Res. 2021;12(1):2267–71.
57. Yang C, Zhao F, Xie C, Zhang Y, Dou Z, Wei X. Community-based group rehabilitation program for stroke patients with dysphagia on quality of life, depression symptoms, and swallowing function: a randomized controlled trial. BMC Geriatr. 2023;23(1):876.
58. Zhang B, Ding P, Hui V, Wong KP, Liu Y, Liu Z, et al. Technology acceptance of the video game-based swallowing function training system among healthcare providers and dysphagia patients: a qualitative study. Digit Health. 2024;10:20552076241284830.
59. Chung HH, Hsieh HF, Jian YJ, Hsueh SC, Liang TY. An integrated rehabilitation APP for stroke patients. In: Proceedings of the 2024 8th international conference on medical and health informatics. New York: Association for Computing Machinery; 2024. p. 127–32. https://doi.org/10.1145/3673971.3673979.
60. Mudumbi S. Poster 160 biofeedback therapy in neurogenic bladder dysfunction. Arch Phys Med Rehabil. 2011;92(10):1737.
61. Abello A, Das AK. Electrical neuromodulation in the management of lower urinary tract dysfunction: evidence, experience and future prospects. Ther Adv Urol. 2018;10(5):165–73.
62. Ma S, Fan X, He Y, Li C, Qu D, Man Y. Effectiveness and safety of digital rectal stimulation and abdominal massage for neurogenic bowel dysfunction in stroke patients: a randomized controlled trial protocol. Trials. 2023;24(1):633.
63. Rapidi CA, Del Popolo G, Spinelli M, Kontaxakis A, Vasilakis R, Sampogna G. Chapter 16 - Telerehabilitation in neurogenic bladder and bowel dysfunction. In: Alexander M, editor. Telerehabilitation. New Delhi: Elsevier; 2022. p. 225–49. Available from: https://www.sciencedirect.com/science/article/pii/B9780323824866000162.
64. Wang X, Xu X, Luo J, Chen Z, Feng S. Effect of app-based audio guidance pelvic floor muscle training on treatment of stress urinary incontinence in primiparas: a randomized controlled trial. Int J Nurs Stud. 2020;104:103527.
65. Candido TA, Ribeiro BM, de Araújo CRC, Pinto R, Resende AP, Pereira-Baldon VS. Effects of tibial and parasacral nerve electrostimulation techniques on women with poststroke overactive bladder: study protocol for a randomized controlled trial. Trials. 2020;21(1):936.
66. Atak Çakir P, Mutluay F, Hanoğlu L, Güzelburç V. Effect of transcutaneous posterior tibial nerve stimulation and repetitive transcranial magnetic stimulation on neurogenic overactive bladder symptoms in female patients with multiple sclerosis: the study protocol of a randomized controlled study. Front Neurol. 2022;13:1011502.
67. Girtner F, Fritsche HM, Zeman F, Huber T, Haider M, Pickl C, et al. Randomized crossover-controlled evaluation of simultaneous bilateral transcutaneous electrostimulation of the posterior Tibial nerve during urodynamic studies in patients with lower urinary tract symptoms. Int Neurourol J. 2021;25(4):337.
68. Laver KE, Adey-Wakeling Z, Crotty M, Lannin NA, George S, Sherrington C. Telerehabilitation services for stroke. Cochrane Database Syst Rev. 2020;1(1):CD010255. https://doi.org/10.1002/14651858.CD010255.pub3.69.

69. Intercollegiate Stroke Working Party. National Clinical Guideline for Stroke. 2023 edition. London: Royal College of Physicians; 2023. Available from: National Clinical Guideline for Stroke website.
70. Kayola G, Mataa MM, Asukile M, Chishimba L, Chomba M, Mortel D, et al. Stroke rehabilitation in low- and middle-income countries: challenges and opportunities. Am J Phys Med Rehabil. 2023;102(2 Suppl 1):S24-S32. https://doi.org/10.1097/PHM.0000000000002128.
71. Niyomyart A, Ruksakulpiwat S, Benjasirisan C, Phianhasin L, Nigussie K, Thorngthip S, et al. Current status of barriers to mHealth access among patients with stroke and steps toward the digital health era: systematic review. JMIR Mhealth Uhealth. 2024;12:e54511. https://doi.org/10.2196/54511.
72. Lindsay MP, Norrving B, Furie KL, Donnan G, Langhorne P, Davis S, on behalf of the Global Stroke Quality and Guidelines Advisory Committee, the Global Stroke Guidelines Working Group, and the Global Stroke Quality Working Group. Global Stroke Guidelines and Action Plan: a road map for quality stroke care. Geneva: World Stroke Organization; 2016. 73.
73. NeuroRehab Directory. Neuro Rehab Products & Equipment [Internet]. 2026 [cited 2026 Apr 10]. Available from: NeuroRehab Directory

Discharge Planning and Transition of Care

10

Kiren Koshy, Veena Babu, Vinicus Viana Abreu, Norhayati Hussein, and P. N. Sylaja

Abstract

Discharge planning is an important component in the care process of patient management, especially for those recovering from a stroke. A structured discharge planning process identifies the type of care a patient with stroke will likely encounter after being discharged from a healthcare facility. This ensures that patients receive full support during their recovery after leaving the hospital. Continuity of care is vital to prevent adverse events and reduce unnecessary hospital readmissions.

Early supported discharge (ESD) is an extension of stroke unit care in the community, wherein the patient is discharged home early, and a multidisciplinary stroke team coordinates the home-based treatment and rehabilitation efforts. Good community support and caregiver motivation, as well as patient responses, are vital for this model. There are significant barriers that contribute to ESD in low- and middle-income countries (LMICs).

Transition of care (ToC) involves ensuring a smooth transition from care in the inpatient setting to the community. A comprehensive discharge plan with community-based rehabilitation efforts can ensure a smooth transition of care. However, significant financial, social, infrastructural, and educational barriers persist in the implementation of a smooth ToC in many LMICs.

K. Koshy · V. Babu · P. N. Sylaja (✉)
Comprehensive Stroke Care Program, Department of Neurology, Sree Chitra Tirunal Institute for Medical Sciences and Technology, Thiruvananthapuram, Kerala, India

V. V. Abreu
Hospital SARAH Kubitschek, Brasilia, Brazil

N. Hussein
Neurological Rehabilitation Unit, Department of Rehabilitation Medicine, Hospital Rehabilitasi Cheras (HRC), Jalan Yaacob Latif, Bandar Tun Razak, Kuala Lumpur, Malaysia

D. B. C. Gandhi, J. D Pandian (eds.), *Stroke Rehabilitation in Low-Resource Settings*, https://doi.org/10.1007/978-981-92-0403-8_10

Keywords

Discharge planning • Transition of care • Early supported discharge

10.1 Executive Summary

Effective discharge planning is essential for an easy transition from hospital to rehabilitation or home. This reduces further complications, shortens hospital stays, and improves patients' quality of life—especially in low- and middle-income countries (LMICs) with limited dedicated stroke centers. Discharge planning ensures a smooth transition for stroke patients across care settings—from hospital to rehabilitation, long-term care, or home. It should address medical, psychosocial, and personal needs through a comprehensive, individualized plan. Effective discharge planning starts at admission, with continuous assessment and support from an interdisciplinary team. This includes early initiation of discharge planning, goal-oriented clinical planning, team coordination, and active patient and family involvement.

Identifying barriers, including lack of caregivers, limited services, cost, or cultural constraints, is a key component of sustainable discharge planning. Evidence shows that support systems like living with someone, being married, and community resources predict a successful return to the home. Despite planning, patients may still face post-discharge concerns such as unaffordable services or information gaps. To mitigate this, continuous patient and caregiver education—reinforced at discharge with written instruction—is vital. Technology-supported tools (audiovisual aids, interactive learning) can further improve preparedness, recovery, and satisfaction.

Several stroke survivors are discharged early with residual disability, and poor post-discharge care increases complications, recurrence, and readmissions. Early supported discharge (ESD) enables survivors to return home earlier with coordinated, home-based rehabilitation by a multidisciplinary team, improving outcomes, satisfaction, and drastically reducing costs. In high-income countries (HICs), ESD is widely implemented, with structured follow-up and ideal case management models that improve functional independence and reduce long-term dependency. In low- and middle-income countries (LMICs), ESD faces barriers such as limited rehabilitation specialists, inadequate services, financial constraints, and caregiver burden. Innovative approaches like Kaifukuki wards, family-led care, tele-rehabilitation, and community-based models have shown improvements, but evidence remains limited. However, there are still several chances in LMICs, like expanding task-shifting models, integrating technology, improving caregiver involvement, and developing structured ESD protocols, to optimize recovery and reduce costs.

Transition of care (ToC) bridges the services from hospital discharge to community reintegration. Challenges in LMICs include limitations on healthcare infrastructure, socioeconomic barriers, health literacy and cultural factors, and psychosocial stressors. Comprehensive discharge planning, utilization of community health workers (CHWs), telehealth integration, caregiver involvement and training, community-based rehabilitation, and policy-level interventions play a leading role in optimizing the transition of care.

In order to address the knowledge gaps, further research is needed in low-resource settings on mHealth for adherence and rehabilitation, integrating mental health into ToC, and reducing disparities in access to care in terms of gender, ethnicity, and socioeconomic status.

10.2 Discharge Planning

Proper discharge planning involves developing an individualized plan to facilitate a patient's transition from a stroke unit to a rehabilitation center or from a rehabilitation center to home, while preventing problems after discharge [1]. Individualized discharge plans can also reduce delays in transition caused by nonmedical factors, help contain medical costs, and improve patient outcomes [1].

Effective discharge planning, including key organizational elements and active patient involvement, is essential for optimal discharge procedures. Appropriate and adequate discharge planning has the potential to provide benefits such as shorter hospital stays, reduced readmission rates, and higher patient satisfaction [2]. In LMICs with limited hospital beds, especially in dedicated stroke centers, this is important so that more patients can benefit from specialized management.

10.3 Principles of Discharge Planning

A discharge plan facilitates smooth transitions for stroke patients in different aspects of care, including from the hospital to home, rehabilitation centers, or long-term care facilities—a process that requires careful and intentional planning. A discharge plan should address all medical, psychosocial, and personal needs that must be managed during a stroke patient's next phase of care.

In order to customize a discharge plan to an individual stroke patient's needs, healthcare professionals must conduct a comprehensive assessment and collaborate with an interdisciplinary stroke rehabilitation team. The ongoing, dynamic process of assessment, treatment, and discharge planning begins at the initial stroke rehabilitation admission and continues throughout the patient's recovery journey.

Effective processes must be established to ensure that follow-up arrangements are made before the patient leaves the healthcare facility, and that any necessary referrals are promptly addressed. Stroke patients and their caregivers should be actively connected in the care transition process starting from the time of treatment. Actively involving stroke patients and their caregivers in the planning process empowers them in self-care and enhances patient satisfaction. This approach prepares them effectively after discharge from the hospital.

The principle of discharge planning for stroke patients should be based on the following guiding principles [3]:

1. Early initiation of discharge planning at admission.
2. Development of a clinical management plan, such as goal setting.
3. Coordination and collaboration among the care team.
4. Active involvement of the patient in the discharge planning process.

Involving the stroke patient and their family early in all phases of the discharge planning process is crucial to ensure seamless continuation of care at home. Potential discharge barriers must be identified and addressed early. These may include obstacles to at-home care, such as a lack of social support or caregivers, availability of specialized services (including transportation and community resources), and personal preferences like service costs or religious and cultural considerations. Patients should be given care providers at their comfort and any cost coverage.

Age, stroke severity at admission, and the availability of a caregiver play important roles in determining the post-rehabilitation discharge planning for individuals with severe stroke. Age and stroke severity should be considered to support appropriate goal setting and discharge planning during inpatient rehabilitation [4]. For individuals with severe stroke, discharge to their home is unlikely without the presence of an available caregiver [5].

During discharge planning, healthcare professionals should evaluate the availability of support at home for stroke patients alongside other outcomes, such as functional status. A review by Chevalley et al. found that living at home, having available support, living with someone, and being married were predictive factors for returning home after inpatient stroke rehabilitation [6]. Therefore, incorporating these factors in the clinical discharge planning process will be essential.

Even when stroke patients overcome discharge planning barriers, they may still face post-discharge challenges at home. According to Temehy et al., common concerns with stroke survivors after discharge from rehabilitation facilities included a lack of readily available information, unsuitable treatment, and inadequate services [7]. For instance, essential clinical services such as nasogastric feeding tube changes or blood pressure monitoring may not be accessible in certain regions.

To ensure stroke patients and their caregivers are well-informed, patient education should be provided while they are in the hospital. At discharge, additional education and written instructions should be given to confirm the patient's understanding of all necessary measures to support their recovery, including medication management and equipment use.

In addition to conventional patient education methods such as booklets, textbooks, and PowerPoint presentations, technology-supported discharge planning can be highly effective. Using audiovisual applications for interactive learning enhances discharge planning, improves patient outcomes, increases family preparedness for discharge, and elevates the overall quality of hospital care [8].

10.4 Early Supported Discharge

Usually, stroke survivors are in the early stages of recovery at the time of discharge, with over three-quarters having significant physical sequelae [9]. Lack of therapeutic and diagnostic expertise, as well as poor focus on early rehabilitation among patients and caregivers, affects the long-term outcome. This results in a higher incidence of complications, stroke recurrence, and an increased readmission rate. Stroke recurrence and readmission can substantially lower the quality of life among stroke survivors. Studies show that there is a significant readmission rate in these patients, ranging from over 15% at 1 month to over 40% over 1 year [10]. Hence, proper planning is required to facilitate an early discharge, while ensuring that readmissions and stroke recurrences are few.

Newer service paradigms offer stroke inpatients an early discharge with home-based rehabilitation. Early supported discharge (ESD) is a model of care that links and allows early discharge, coupled with early initiation of focused rehabilitation. ESD services should complement, not replace, inpatient stroke unit care. A multidisciplinary team of therapists, nurses, and doctors coordinates the patient's discharge and provides home-based rehabilitation. ESD is most appropriate for patients with minor or intermediate grades of disability. It can improve patient outcomes and patient satisfaction scores while limiting cost. As part of ESD, domiciliary support is offered from a hospital-based team or a dedicated outpatient team [11]. Functional outcomes, especially independence of activities of daily living (ADL), are significantly reduced with ESD, especially in patients with limited disability [12]. The duration of ESD depends on the patient's disability status.

At present, early supported discharge is not an integral part of stroke care in most low- and middle-income countries (LMICs). The cost of the ESD program was significantly lower than that of hospital-based treatment [13]. Implementation of this program in LMICs has enormous potential in terms of optimizing outcomes and cutting costs.

10.4.1 ESD in High-Income Countries

Many high-income countries, including the United States, Canada, Scotland, and Australia, recommend early supported discharge services as a rehabilitation mechanism for post-acute stroke care [14–17]. The National Stroke Foundation of Australia has introduced a concept of a case management model after discharge, wherein the case managers recognize and manage post-stroke symptoms like depression and help to coordinate any interventions through medical practitioners [17]. They also advocate a well-documented post-discharge care plan with follow-up by a member of the stroke team within 3 months and then at 6- and 12-month post-discharge [17]. ESD with further management either in a day unit or in the survivors' own home improved the functional independence score, trunk control, and walking among the survivors [18, 19].

10.4.2 ESD in Low- and Middle-Income Countries

ESD in LMICs needs a structured plan that takes into account the strengths and challenges unique to each country. The nonavailability of post-discharge inpatient rehabilitation facilities, physical medicine and rehabilitation doctors, speech therapists, and occupational therapists is the prime issue in LMICs. Moreover, care from an informal caregiver adds to the delayed recovery of stroke survivors. The majority of the survivors and their families lack the capacity to purchase ambulatory aids or wheelchairs. Reduced transportation, as in many LMICs, makes the survivor depend on the shoulders of the caregivers, which in turn deteriorates the caregiver's health [20, 21].

A Kaifukuki rehabilitation ward is a facility for focused rehabilitation by a multidisciplinary team, with the aim of early transfer to home-based rehabilitation [22]. Acute stroke unit care followed by rapid transfer to a rehabilitation ward with early supported discharge decreases mortality and complications after acute ischemic stroke [23]. A family-led stroke rehabilitation model with a concept of task shifting to family members did not show any improvement in dependency or rehospitalization rates with this model [24].

Stroke rehabilitation could be improved by encouraging self-management for the physical and psychological challenges in the recovery trajectory. Physical and psychological disability, occupational limitations, and paucity of social support remain significant challenges. Appropriate self-management of such challenges has the potential to transform clinical outcomes. This requires conscious effort by survivors themselves, and limited evidence exists for its efficacy at present [25]. Dedicated stroke rehabilitation services are often lacking in LMICS, often due to a lack of funds and logistical support [26]. Tele-rehabilitation and technology-assisted rehabilitation could overcome these challenges in the future.

10.4.3 Challenges/Barriers for ESD in LMICs

The lack of neurologists and specialist stroke care professionals has created a need for alternative methods of stroke care in the community. Although the nonspecialist model of stroke care is cost-effective and culturally acceptable in LMICs, its impact on a large scale has not been studied [27]. Distance to the nearest health facility with physiotherapy services and uncontrolled risk factors like hypertension or diabetes, which may need specialist care, are also challenges for early supported discharge for stroke survivors. Healthcare workers from HICs are in a better position to prioritize, facilitate, and implement specialized stroke rehabilitation than their counterparts in LMICs.

Though several middle-income countries like South Africa and the Philippines have adopted standardized protocols for early supported discharge, less than half of the LMICs have standardized guidelines or protocols for early supported discharge [28]. Stroke units (defined as standardized systems of primarily nursing-based care for patients with stroke) are associated with improved functional outcomes as well

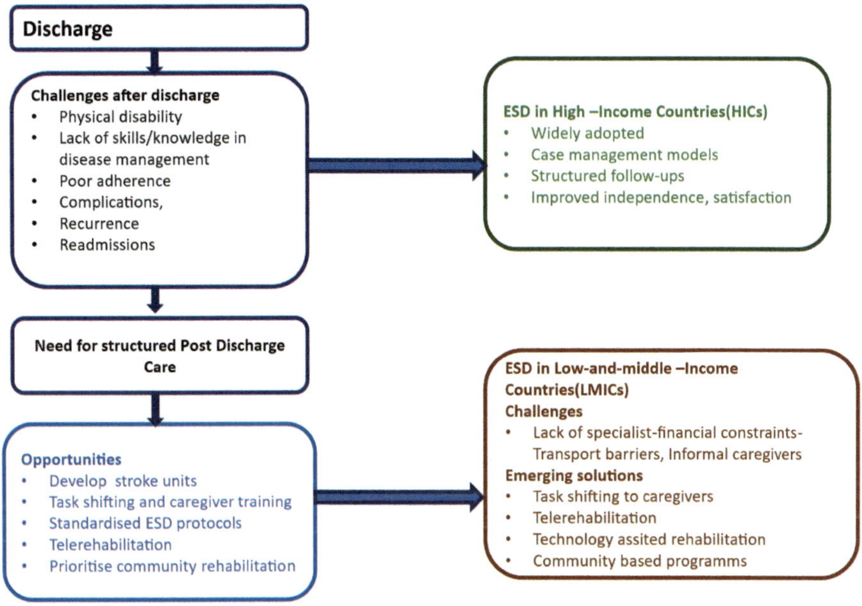

Fig. 10.1 Framework of early supported discharge (ESD) in stroke care: challenges, contextual differences, and opportunities

as reduced death and dependency rates in acute stroke. A lack of specialized stroke units is seen in most of the LMICs, especially in sub-Saharan Africa [29].

10.4.4 Opportunities

Task shifting training, increasing interventions by rehabilitation specialists, development of structured stroke units with a focus on ESD, community-based rehabilitation programs, and caregiver-led rehabilitation should be prioritized for ESD in LMICs. The framework of ESD is given in Fig. 10.1.

10.5 Transition of Care

Transition of care (ToC) represents a pivotal component in stroke management and rehabilitation, especially in low-resource settings where disparities in healthcare infrastructure pose significant challenges. Ensuring continuity of care from hospital discharge to the community is essential to reduce the risk of complications, optimize recovery, and improve the quality of life for stroke survivors. This section explores the challenges and opportunities in facilitating effective transitions, with a focus on evidence-based interventions tailored to resource-limited contexts.

10.5.1 Challenges in Low-Resource Settings

1. Healthcare infrastructure limitations.
   (a) The availability of post-acute rehabilitation services, including outpatient physical therapy and occupational therapy, is often insufficient in low-resource environments. Rehabilitation centers, where they exist, are frequently concentrated in urban areas, leaving rural populations underserved [30, 31].
   (b) Equipment shortages and inadequate training for healthcare professionals exacerbate these challenges, limiting the ability to provide tailored rehabilitation plans.
2. Socioeconomic barriers.
   (a) Financial constraints hinder access to rehabilitation and follow-up care. Many patients cannot afford medications or transportation to distant clinics, resulting in suboptimal adherence to prescribed therapies [32].
   (b) Employment loss due to disability further compounds economic strain, particularly in communities where social safety nets are limited or nonexistent [33].
3. Health literacy and cultural factors.
   (a) Inadequate health literacy complicates patient and caregiver understanding of discharge instructions, medication regimens, and the importance of follow-up care [34]. This issue is compounded by cultural beliefs that may deprioritize rehabilitation.
   (b) Language barriers between patients and healthcare providers can also impede effective communication, necessitating culturally sensitive approaches.
4. Psychosocial stressors.
   (a) Stroke survivors and their families often experience significant psychological and emotional stress during the transition period. This burden is heightened in low-resource settings, where mental health services are frequently unavailable [35].

10.5.2 Strategies for Optimizing Transitions

1. Comprehensive discharge planning
   (a) Effective ToC begins at the hospital. Evidence supports the use of structured discharge planning to ensure that stroke survivors leave the hospital with clear, actionable instructions for follow-up care, including rehabilitation schedules, medication regimens, and red flags for complications [30] (Table 10.1).
   (b) Discharge summaries should be concise, standardized, and designed for low health literacy levels to improve patient and caregiver understanding [36].
   (c) Refer to the World Stroke Organization's post-stroke discharge checklists to streamline components of assessment (Table 10.1)

Table 10.1 Model discharge plan/checklist

Domain	Details to document/checklist
Patient details	Name: ______ • age: ______ • hospital no: ______ • diagnosis: ________________
Functional status	GCS: ______ • modified Rankin score: ______
Dysphagia management	Feeding mode: ☐ full oral ☐ modified oral ☐ NGT ☐ repeat swallow assessment planned NGT removal timing_________ ☐ explained precautions ☐ strategies for medication administration if dysphagia is present.
Bladder management	☐ indwelling catheter Insertion:_________ Change time:________ Removal plan:___________ ☐ condom catheter ☐ conscious voiding
Special bedding and positioning	☐ air/water bed ☐ head-end elevation ☐ position changes ☐ bed sore prevention/treatment
Physiotherapy	☐ active physiotherapy Plan and personnel/setting to be used________ ☐ passive physiotherapy Plan and personnel/setting to be used_________ ☐ gait training and precautions against falls
Speech	☐ speech assessment summary documented ☐ therapy plan provided
Medical management	☐ current medication list ☐ side effects/interaction explained Contact for medicine related issues:____________ Local pharmacy details:_________________________
Comorbidities	☐ diabetes ☐ hypertension ☐ dyslipidemia ☐ CADmonitoring frequency: ______☐ done with local physician coordination ☐ potential issues/emergency measures explained (e.g., hypoglycemia)
Follow-up	Local physician: Next visit on ______stroke center: Next visit on ______
Cognitive issues/ psychiatric issues	☐ screening for post-stroke depression or dementia done ☐ medical management (if appropriate) done ☐ cognitive behavior therapy/cognitive retraining done ☐ supervised medication administration (if indicated)
Barriers along with strategies to overcome	Financial______________ Economic______________ Logistic_______________
Patient/caregiver perspectives	
Emergency contact number	

(https://world-stroke-academy.org/media/uploads/2022/09/WSO-Stroke-Rehabilitation-Checklist.pdf; https://www.world-stroke.org/assets/downloads/psc-us-version-wso-livery-03-19-13.pdf).

2. Utilization of community health workers (CHWs)
    (a) CHWs play a transformative role in bridging gaps between formal healthcare services and community-based support. In resource-limited contexts, CHWs provide home visits, facilitate adherence to treatment protocols, and deliver culturally appropriate education [24, 31].

        For example, India has a cadre of Accredited Social Health Activists (ASHAs) who provide community-based health education and basic health services.
    (b) Training CHWs in basic rehabilitation techniques and psychosocial support ensures they can address common post-stroke complications effectively.
3. Telehealth integration
    (a) Telehealth platforms are increasingly recognized as effective tools for providing follow-up care in remote areas. Tele-rehabilitation, involving virtual sessions with physiotherapists and neurologists, has shown promise in improving motor and cognitive outcomes in stroke survivors [30].
    (b) Despite initial setup costs, telehealth can be cost-effective in the long term, reducing transportation expenses and improving access to care.
4. Caregiver involvement and training
    (a) Caregivers are central to the recovery journey. Empowering them through structured training programs that cover physical therapy exercises, medication administration, and signs of stroke recurrence significantly reduces caregiver burden and enhances patient outcomes [31, 32].
    (b) Peer support groups for caregivers also provide emotional support and shared learning opportunities, which are particularly beneficial in low-resource contexts.
5. Community-based rehabilitation (CBR)
    (a) CBR programs integrate health, education, and social services at the community level. In low-resource settings, these programs have been shown to improve functional recovery and reintegration into society for stroke survivors [33].
    (b) Local governments and NGOs play a crucial role in running these programs, leveraging local knowledge and resources to maximize their impact.
6. Policy and system-level interventions
    (a) Health policies that prioritize stroke rehabilitation and allocate funding for infrastructure development are fundamental. Policymakers must focus on creating multidisciplinary stroke care pathways that extend beyond acute care settings [36].
    (b) Partnerships with international organizations can provide technical assistance and funding to strengthen local capacities.

10.6 A Case Study: Nigeria's Stroke Transition Program

In Nigeria, a novel ToC program demonstrated the feasibility of low-cost interventions in resource-limited environments. The program combined structured discharge planning with CHW involvement and telehealth check-ins, resulting in a 25% reduction in hospital readmissions and improved patient satisfaction scores over 12 months [31].

10.7 Addressing Gaps in Knowledge

Despite progress, further research is needed to tailor ToC strategies to diverse low-resource settings. Key areas for exploration include:

- The role of mobile health applications in supporting medication adherence and rehabilitation activities.
- Effective models for integrating mental health services into ToC frameworks.
- Strategies to address disparities in access to care based on gender, ethnicity, and socioeconomic status [35].

10.8 Conclusion

Discharge planning is essential to ensure optimal use of resources in the LMIC setting, while ensuring patient safety and well-being. A well-written discharge plan has to be prepared by the treating team early in the patient's inpatient journey. Early supported discharge is appropriate for patients with mild to moderate deficits and good family or community support. Transition of care is a vital phase in stroke recovery, with profound implications for patient outcomes and healthcare systems. In low-resource settings, addressing barriers such as limited access to rehabilitation, socioeconomic constraints, and low health literacy is critical. By implementing evidence-based strategies, including comprehensive discharge planning, CHW involvement, and telehealth interventions, it is possible to improve the quality and continuity of care. Collaborative efforts among healthcare providers, policymakers, and communities are essential to bridging gaps and ensuring equitable access to post-stroke care. Further research is needed in the fields of caregiver-led rehabilitation and technology-aided rehabilitation.

What Does Evidence Say

- Early supported discharge should be offered to stroke survivors with mild to moderate disability where coordinated home-based services are available.
- Home-based rehabilitation may be the preferred model in the community; if not available, center-based rehabilitation should be provided.
- A comprehensive, individualized discharge care plan should be developed collaboratively with the stroke survivor and their caregiver prior to discharge.
- Stroke survivors and their families/carers should receive tailored information in appropriate language and formats, delivered at multiple stages of recovery.
- Information delivery should involve active engagement, with opportunities for follow-up, clarification, and reinforcement.
- Caregivers should receive structured, individualized training before discharge from the relevant interdisciplinary team members.
- Training may include personal care, communication techniques, safe handling, prevention strategies, swallowing and dietary guidance, and management of behavioral or psychosocial issues.
- Stroke survivors should be offered self-management interventions within the first 4 months after discharge, with approaches such as "Take Charge After Stroke" having strong evidence.
- Patients and carers should be informed about and supported to access generic or stroke-specific self-management programs, with collaborative care plans used where appropriate.
- Stroke survivors and carers should be informed about local stroke support groups and other peer support options before discharge and during community reintegration.
- Caregivers should receive continuous, tailored information and support throughout recovery, including access to healthcare professionals, understanding care plans, and navigating community services.
- Support can be delivered in person, via phone, or on digital platforms.
- Carers should be offered psychological and social support across the recovery continuum, including strategies for problem-solving, coping, and stress management.
- Support should also address relationship dynamics to maintain the well-being of both the caregiver and the stroke survivor.

References

1. Mistiaen P, Francke AL, Poot E. Interventions aimed at reducing problems in adult patients discharged from hospital to home: a systematic meta-review. BMC Health Serv Res. 2007; https://doi.org/10.1186/1472-6963-7-47.
2. Gonçalves-Bradley DC, Lannin NA, Clemson LM, Cameron ID, Shepperd S. Discharge planning from hospital. Cochrane Database Syst Rev. 2016;2016(1):CD000313. https://doi.org/10.1002/14651858.CD000313.pub5. PMID: 26816297; PMCID: PMC7073416.
3. Heath H, Sturdy D, Cheesely A. Discharge planning: a summary of the department of health's guidance: ready to go? Planning the discharge and the transfer of patients from hospital and intermediate care. Harrow: RCN Publishing Company; 2010.
4. Burton JK, Ferguson EEC, Barugh AJ, Walesby KE, MacLullich AMJ, Shenkin SD, Quinn TJ. Predicting discharge to institutional long-term care after stroke: a systematic review and metaanalysis. J Am Geriatr Soc. 2018;66(1):161–9. https://doi.org/10.1111/jgs.15101. Epub 2017 Oct 9. PMID: 28991368; PMCID: PMC5813141.
5. Pereira S, Foley N, Salter K, McClure JA, Meyer M, Brown J, Teasell R. Discharge destination of individuals with severe stroke undergoing rehabilitation: a predictive model. Disabil Rehabil. 2014;36(9):727–31.
6. Chevalley O, Truijen S, Saeys W, Opsommer E. Socio-environmental predictive factors for discharge destination after inpatient rehabilitation in patients with stroke: a systematic review and meta-analysis. Disabil Rehabil. 2022;44(18):4974–85. https://doi.org/10.1080/09638288.2021.1923838. Epub 2021 May 18. PMID: 34004119.
7. Temehy B, Rosewilliam S, Alvey G, Soundy A. Exploring stroke patients' needs after discharge from rehabilitation centers: meta-ethnography. Behav Sci. 2022;12(10):404. https://doi.org/10.3390/bs12100404. PMID: 36285973; PMCID: PMC9598696.
8. Rosadi M, Arofiati F. Development of discharge planning for stroke patients: systematic review. Folia Medica Indonesiana. 2023;59:400–9. https://doi.org/10.20473/fmi.v59i4.42819.
9. Chen WW, Gao RL, Liu LS, Zhu ML, Wang W, Wang YJ, et al. China cardiovascular diseases report 2015: a summary. J Geriatr Cardiol. 2017;14(1):1.
10. Zhong W, Geng N, Wang P, Li Z, Cao L. Prevalence, causes and risk factors of hospital readmissions after acute stroke and transient ischemic attack: a systematic review and meta-analysis. Neurol Sci. 2016;37(8):1195–202. https://doi.org/10.1007/s10072-016-2570-5. Epub 2016 Apr 29.
11. Fearon P, Langhorne P, Early Supported Discharge Trialists. Services for reducing duration of hospital care for acute stroke patients. Cochrane Database Syst Rev. 2012;9:CD000443. https://doi.org/10.1002/14651858.CD000443.pub3.
12. Fisher RJ, Cobley CS, Potgieter I, Moody A, Nouri F, Gaynor C, Byrne A, Walker MF. Is stroke early supported discharge still effective in practice? A prospective comparative study. Clin Rehabil. 2016;30(3):268–76. https://doi.org/10.1177/0269215515578697. Epub 2015 Mar 31.
13. Langhorne P, Baylan S. Early supported discharge trialists. Early supported discharge services for people with acute stroke. Cochrane Database Syst Rev. 2017;7(7):CD000443. https://doi.org/10.1002/14651858.CD000443.pub4. PMID: 28703869; PMCID: PMC6483472.
14. Sall J, Eapen BC, Tran JE, Bowles AO, Bursaw A, Rodgers ME. The management of stroke rehabilitation: a synopsis of the 2019 U.S. Department of Veterans Affairs and U.S. Department of Defense clinical practice guideline. Ann Intern Med. 2019;171(12):916–24.
15. Mountain A, Patrice Lindsay M, Teasell R, Salbach NM, de Jong A, Foley N, et al. Canadian stroke best practice recommendations: rehabilitation, recovery, and community participation following stroke. Part two: transitions and community participation following stroke. Int J Stroke. 2020;15(7):789–806.
16. Smith LN, James R, Barber M, Ramsay S, Gillespie D, Chung C, et al. Rehabilitation of patients with stroke: summary of SIGN guidance. BMJ. 2010;340:c2845.

17. Middleton S, Rowley J, Hillege S, Hill K, Churilov L, Allnutt N. Clinical guidelines for acute stroke management: which recommendations should remain consensus-based? J Vasc Nurs. 2013;31(2):72–83.
18. Santana S, Rente J, Neves C, Redondo P, Szczygiel N, Larsen T, Jepsen B, Langhorne P. Early home-supported discharge for patients with stroke in Portugal: a randomised controlled trial. Clin Rehabil. 2017;31(2):197–206. https://doi.org/10.1177/0269215515627282. Epub 2016 Jul 10. PMID: 26837431; PMCID: PMC5302124.
19. Gjelsvik BE, Hofstad H, Smedal T, Eide GE, Næss H, Skouen JS, Frisk B, Daltveit S, Strand LI. Balance and walking after three different models of stroke rehabilitation: early supported discharge in a day unit or at home, and traditional treatment (control). BMJ Open. 2014;4(5):e004358. https://doi.org/10.1136/bmjopen-2013-004358. PMID: 24833680; PMCID: PMC4025466.
20. Kayola G, Mataa MM, Asukile M, Chishimba L, Chomba M, Mortel D, Nutakki A, Zimba S, Saylor D. Stroke rehabilitation in low- and middle-income countries: challenges and opportunities. Am J Phys Med Rehabil. 2023;102(2S Suppl 1):S24–32. https://doi.org/10.1097/PHM.0000000000002128. PMID: 36634327; PMCID: PMC9846582.
21. Mkoba EM, Sundelin G, Sahlen KG, Sörlin A. The characteristics of stroke and its rehabilitation in northern Tanzania. Glob Health Action. 2021;14(1):1927507. https://doi.org/10.1080/16549716.2021.1927507. PMID: 34340643; PMCID: PMC8344769.
22. Katsuki M, Narita N, Ozaki D, Sato Y, Jia W, Nishizawa T, Kochi R, Sato K, Kawamura K, Ishida N, Watanabe O, Cai S, Shimabukuro S, Yasuda I, Kinjo K, Yokota K. Deep learning-based functional independence measure score prediction after stroke in Kaifukuki (convalescent) rehabilitation ward annexed to acute care hospital. Cureus. 2021;13(7):e16588. https://doi.org/10.7759/cureus.16588. PMID: 34466308; PMCID: PMC8396410.
23. Suwanwela NC, Eusattasak N, Phanthumchinda K, Piravej K, Locharoenkul C. Combination of acute stroke unit and short-term stroke ward with early supported discharge decreases mortality and complications after acute ischemic stroke. J Med Assoc Thail. 2007;90(6):1089–96.
24. ATTEND Collaborative Group. Family-led rehabilitation after stroke in India (ATTEND): a randomised controlled trial. Lancet. 2017;390(10094):588–99. https://doi.org/10.1016/S0140-6736(17)31447-2.
25. Lennon S, McKenna S, Jones F. Self-management programmes for people post stroke: a systematic review. Clin Rehabil. 2013;27(10):867–78. https://doi.org/10.1177/0269215513481045. Epub 2013 Mar 29.
26. Linder SM, Rosenfeldt AB, Bay RC, Sahu K, Wolf SL, Alberts JL. Improving quality of life and depression after stroke through telerehabilitation. Am J Occup Ther. 2015;69(2):6902290020p1-10. https://doi.org/10.5014/ajot.2015.014498. PMID: 26122686; PMCID: PMC4480056.
27. Sebastian IA, Demers M, Yeghiazaryan N, Asyraf WZW, Nasreldein A, Gopaul U, et al. Establishing organized stroke care in low- and middle-income countries: from training of non-specialist to implementation. J Stroke Med. 2023;6(1):46–51.
28. Owolabi MO, Thrift AG, Martins S, Johnson W, Pandian J, Abd-Allah F, et al. The state of stroke services across the globe: report of world stroke organization-World Health Organization surveys. Int J Stroke. 2021;16(8):889–901.
29. Stroke Unit Trialists' Collaboration. Organised inpatient (stroke unit) care for stroke. Cochrane Database Syst Rev. 2013;2013(9):CD000197. https://doi.org/10.1002/14651858.CD000197.pub3.
30. Kernan WN, Viera AJ, Billinger SA, Bravata DM, Stark SL, Kasner SE, Kuritzky L, Towfighi A, American Heart Association Stroke Council, Council on Arteriosclerosis, Thrombosis and Vascular Biology; Council on Cardiovascular Radiology and Intervention; and Council on Peripheral Vascular Disease. Primary care of adult patients after stroke: a scientific statement from the American Heart Association/American Stroke Association. Stroke. 2021;52(9):e558–71. https://doi.org/10.1161/STR.0000000000000382. Epub 2021 Jul 15.
31. Reshetnyak E, Ntamatungiro M, Pinheiro LC, Howard VJ, Carson AP, Martin KD, Safford MM. Impact of multiple social determinants of health on incident stroke. Stroke.

2020;51(8):2445–53. https://doi.org/10.1161/STROKEAHA.120.028530. Epub 2020 Jul 16. PMID: 32673521; PMCID: PMC9264323.
32. JNJ B, Zhao Y, Burns S, Magwood G, Adams R, Sims-Robinson C, Lackland DT, WISSDOM Research Center Study Group. Racial disparities in stroke recovery persistence in the post-acute stroke recovery phase: evidence from the health and retirement study. Ethn Dis. 2020;30(2):339–48. https://doi.org/10.18865/ed.30.2.339. PMID: 32346280; PMCID: PMC7186057.
33. Sall J, Eapen BC, Tran JE, Bowles AO, Bursaw A, Rodgers ME. The management of stroke rehabilitation: a synopsis of the 2019 U.S. Department of Veterans Affairs and U.S. Department of Defense clinical practice guideline. Ann Intern Med. 2019;171(12):916–24. https://doi.org/10.7326/M19-1695. Epub 2019 Nov 19.
34. Bernheisel CR, Schlaudecker JD, Leopold K. Subacute management of ischemic stroke. Am Fam Physician. 2011;84(12):1383–8.
35. Judd SE, Gutiérrez OM, Newby PK, Howard G, Howard VJ, Locher JL, Kissela BM, Shikany JM. Dietary patterns are associated with incident stroke and contribute to excess risk of stroke in black Americans. Stroke. 2013;44(12):3305–11. https://doi.org/10.1161/STROKEAHA.113.002636. Epub 2013 Oct 24. PMID: 24159061; PMCID: PMC3898713.
36. Tsao CW, Aday AW, Almarzooq ZI, Alonso A, Beaton AZ, Bittencourt MS, Boehme AK, Buxton AE, Carson AP, Commodore-Mensah Y, Elkind MSV, Evenson KR, Eze-Nliam C, Ferguson JF, Generoso G, Ho JE, Kalani R, Khan SS, Kissela BM, Knutson KL, Levine DA, Lewis TT, Liu J, Loop MS, Ma J, Mussolino ME, Navaneethan SD, Perak AM, Poudel R, Rezk-Hanna M, Roth GA, Schroeder EB, Shah SH, Thacker EL, VanWagner LB, Virani SS, Voecks JH, Wang NY, Yaffe K, Martin SS. Heart disease and stroke Statistics-2022 update: a report from the American Heart Association. Circulation. 2022;145(8):e153–639. https://doi.org/10.1161/CIR.0000000000001052.

Models of Stroke Rehabilitation in Low-Resource Settings: The Ground Realities

11

Norhayati Hussein, Jussara Almeida de Oliveira Baggio, and Sureshkumar Kamalakannan

Abstract

Pragmatic stroke rehabilitation models are essential in low-resource settings due to the unique challenges these environments present. Limited access to stroke rehabilitation is a significant issue, often stemming from the shortage of trained healthcare professionals and the lack of rehabilitation facilities equipped with specialized assessment tools and therapeutic equipment. Additionally, financial constraints result in less affordable, high-quality, and long-term rehabilitation services for many stroke survivors. Geographic and locational barriers also pose substantial challenges that must be addressed to ensure optimal access to stroke rehabilitation. Cultural and social factors, coupled with low awareness of stroke prevention and rehabilitation, exacerbate the difficulties faced in these settings. This chapter delves into the range of stroke rehabilitation models that leverage available resources suitable for low-resource settings, examining their benefits and the real-world challenges and opportunities encountered in the implementation.

N. Hussein (✉)
Neurological Rehabilitation Unit, Department of Rehabilitation Medicine, Hospital Rehabilitasi Cheras (HRC), Jalan Yaacob Latif, Bandar Tun Razak, Kuala Lumpur, Malaysia
e-mail: yati.hussein@moh.gov.my

J. A. de Oliveira Baggio
Medical Sciences and Nursing Complex – CCME, Federal University of Alagoas, Arapiraca, Alagoas, Brazil

S. Kamalakannan
Faculty of Health and Life Sciences, Department of Social Work, Education, and Community Wellbeing, Northumbria University, Newcastle upon Tyne, UK

D. B. C. Gandhi, J. D Pandian (eds.), *Stroke Rehabilitation in Low-Resource Settings*, https://doi.org/10.1007/978-981-92-0403-8_11

Keywords

Stroke rehabilitation models • Low resource • Access • Barriers

Learning Objectives

- To understand the implementation determinants of stroke rehabilitation models.
- To identify the various stroke rehabilitation models suitable for implementation in low-resource settings.
- To identify opportunities and solutions to implement the stroke rehabilitation models via real-world case illustration.

11.1 Chapter Outline

This chapter presents a global situational analysis of stroke rehabilitation, emphasizing service disparities in low- and middle-income countries. It examines the key barriers to implementation, such as access, cost, workforce gaps, cultural factors, and evidence translation. Four practical models—community-based rehabilitation, task-shifting, telerehabilitation, and self-management are discussed as solutions for resource-limited settings. Case studies from Brazil, Malaysia, and India illustrate how these models have been adapted and implemented. The chapter concludes with strategies to enhance rehabilitation systems through community involvement, policy reform, capacity building, and innovation.

11.2 Introduction and Situational Analysis of Stroke Rehabilitation Across the Globe

Across the globe, the low- and middle-income countries (LMIC) have far fewer rehabilitation services than recommended. The regions of Latin America, South Asia, and sub-Saharan Africa stand out as having the lowest availability of services [1]. For instance, in Latin America, the main barriers to accessing rehabilitation include a lack of infrastructure, a lack of public health system coverage, a shortage of rehabilitation professionals, and the distance to outpatient rehabilitation services [2].

The shortage of trained professionals is a frequently reported issue across different parts of the world. In the Southeast Asia Region (SEAR), there is a significant lack of professionals, with countries such as Bhutan having only two speech therapists and Timor-Leste having just two physiotherapists [3]. In Africa, 11 countries have speech therapists, though they are accessible only in two sites [4]. In Brazil, until March 2024, there were 55,045 speech therapists, a ratio of 0.25 per 1000 inhabitants, with most concentrated in the Southeast region of the country [5].

Additionally, there is a notable discrepancy between the provision of rehabilitation services during the hospital phase compared to the posthospital discharge phase. In LMICs, inpatient rehabilitation has been reported as considerably well-available (80.95%), while community rehabilitation programs were seen as poorly available (7.84%), and outpatient rehabilitation services were deemed acceptable (42.86%) [1]. These figures highlight the inequality in stroke care investments, which impacts positive functional outcomes.

11.3 Implementation Determinants of Stroke Rehabilitation in Low-Resource Settings.

Multiple and concurrent factors contribute to the effective implementation of stroke rehabilitation. The implementation determinants most often implicated in the challenges of implementing successful rehabilitation models are described below [6–8].

11.3.1 Accessibility

In low-resource settings, geographic locations substantially affect access to rehabilitation care by making it difficult for stroke survivors to access rehabilitation services [7]. The geographic-related barriers include long travel distances to healthcare facilities and a lack of transportation options. This ultimately leads to health disparities in stroke rehabilitation based on geographic locations [7]. Economic barriers, such as the cost of travel and treatment, also play a significant role.

11.3.2 Economic and Financial Challenges

Financial constraints affect both the availability of rehabilitation services and the ability of patients to afford care. Limited funding for healthcare infrastructure and services can result in inadequate facilities and equipment [6]. Additionally, patients may face difficulties financing rehabilitation services, leading to delayed or incomplete rehabilitation.

11.3.3 Rehabilitation Skilled Workforce

There is often a shortage of trained healthcare professionals in low-resource settings. This shortage leads to ineffective quality of rehabilitation service provision and leads to long waiting times for patients [6, 7]. Upskilling the healthcare providers and retaining skilled professionals is a significant challenge.

11.3.4 Cultural Beliefs and Local Practices

Cultural beliefs and local practices strongly influence how rehabilitation is understood and accessed. Misconceptions about stroke and its treatment may cause hesitation or avoidance. Bridging these gaps through education and community engagement is key to promoting acceptance.

11.3.5 Stroke Rehabilitation Education and Awareness

Limited awareness among patients and caregivers about the benefits of rehabilitation and available services can hinder the implementation of rehabilitation models. Educational campaigns and community outreach programs can help increase awareness and encourage the use of rehabilitation services [7, 8].

11.3.6 Rehabilitation Policy and Funding

To implement effective rehabilitation models, supportive policies, and sufficient funding are crucial. In many low-resource settings, there's a noticeable lack of policies that prioritize stroke rehabilitation and allocate the necessary funds for these services. It's important to advocate for policy changes and increased funding to enhance stroke care [7, 8].

11.3.7 Knowledge-to-Practice

Another important barrier to stroke rehabilitation is the translation of evidence-based practice into clinical practice. There is a high priority for developing Clinical Practice Guidelines specific to low-resource settings [9, 10], but several barriers to implementation persist, such as a lack of organizational support processes, time, resources, and insufficient professional knowledge and skills [11]. Therefore, training professionals through remote learning strategies, short exchange programs, and hands-on workshops is essential for the implementation of evidence-based practices [12].

11.4 Stroke Rehabilitation Models and the Ground Realities

11.4.1 Community-Based Rehabilitation

Community-based rehabilitation (CBR) is an important and widely used model for providing stroke rehabilitation, particularly in low-resource settings [13]. CBR was introduced as a global strategy to enhance access to rehabilitation services for persons with disabilities worldwide [14]. It is considered an effective model for improving access to stroke care, care efficiency, coordination, and continuity, especially in low-resource settings.

Two key aspects define the CBR structure [15]. Firstly, CBR must operate at the community level to empower individuals with disabilities. Secondly, the resources for CBR should be optimized from those available within the local community. These two components are essential for achieving the goals envisioned by the CBR model [16]. By focusing on community involvement and utilizing local resources, CBR is expected to provide sustainable and effective rehabilitation services to stroke survivors, even in low-resource settings. However, when it comes to stroke rehabilitation, implementing CBR is more complex and challenging compared to CBR implementation for other kinds of disabilities.

The key challenges arise from diverse reasons. The primary reason for the lack of implementation of stroke care in the community is the nonavailability of a system or pathway in place for stroke care in the community [17]. In many LRS, primary care does not include rehabilitation, let alone stroke care. Whatever is available for stroke rehabilitation is accessible at tertiary care facilities managed by the private sector, which are located in urban areas and predominantly led by one or two rehabilitation-related specialists, like a physiotherapist and/or an occupational therapist [18]. There are no organized pathways for different aspects of stroke care, like a hyperacute stroke unit, acute, or chronic care, even in these facilities and settings [19]. Given the dearth of resources, the quality and effectiveness of stroke care available in the communities of any LRS is compromised [20].

Low-resource settings exist globally and it does not just refer to low- and middle-income countries (LMICs). However, the approach to the provision of community-based stroke care differs between LRS in LMICs compared to LRS in high-income countries (HICs) [21]. The primary difference is the existence of an organized pathway for stroke rehabilitation, even at the level of the community, that is accessible free of cost in most HICs. These kinds of services are not available and affordable even within most of the private tertiary hospitals in LMICs. Hence, the burden of care and rehabilitation falls on the families of the stroke survivors, who become informal caregivers for the stroke survivor and take up the responsibility and roles of the stroke survivors.

11.4.2 Task-Shifting Models

The task-shifting model in stroke rehabilitation involves delegating specific tasks from highly specialized healthcare professionals to less specialized healthcare personnel [22]. The task-shifting model is practical and useful in low-resource settings where there is a shortage of specialized healthcare providers [22, 23]. Task shifting is a process involving moving and shifting appropriate tasks to healthcare providers, which requires shorter training duration and fewer qualifications [22]. The task-shifting model can reduce costs and make rehabilitation services more accessible to a larger number of stroke survivors [22, 23].

The key aspects of the task-shifting models are training and education, supervision and support, task allocation, and community involvement [22, 23].

1. Training and education: Training focused on basic rehabilitation techniques, including hemiplegic care, personal care, communication, and feeding techniques. The training sessions shall also include how to monitor progress and detect common stroke-related complications.
2. Supervision and support: This aspect involves the active presence of the professionals to provide ongoing supervision and support to the less qualified healthcare providers.
3. Task allocation: Identified appropriate tasks are allocated or delegated based on the skills and competencies of the healthcare providers. Specific tasks that are commonly performed by professionals can be delegated to another group of healthcare providers. The delegation allows the professionals more time in managing the more complex stroke patients.
4. Community involvement: The task-shifting model often involves the community in the rehabilitation process. Community members can be trained to assist with certain tasks, providing additional support and resources to stroke survivors.

With the key aspects listed above, the concept of task shifting is viewed as a potentially suitable alternative to conventional rehabilitation demands in low- and middle-income countries, most of which are likely to have low resources. However, the reality is not aligned with the expected potential. Previous studies did not demonstrate significant gains in functionality and did not provide sufficient evidence to fully justify the adoption of the task-shifting model in stroke [22].

The ATTEND trial in India was designed to evaluate the effectiveness of a family-led stroke rehabilitation model in improving clinical outcomes [24]. The ATTEND investigators concluded that the model was not superior to usual care concerning critical outcomes such as mortality, dependency, and rehospitalization, suggesting that it might not be the most efficient use of limited resources [24]. On a similar note, the TRACS trial found no significant difference in stroke patients' recovery, mood, and quality of life, nor in caregivers' burden and perceived cost-effectiveness of stroke caregiver training programs [25]. The findings indicated no significant difference compared to usual care, highlighting the need for more effective interventions [25].

It is apparent that the overall success of task shifting in stroke rehabilitation ultimately depends on appropriate task allocation, proper training, structured supervision, and continual support for the less specialized healthcare providers to ensure quality stroke rehabilitation care can be rendered in the best manner with fewer resources.

11.4.3 Telerehabilitation

Telerehabilitation involves providing and overseeing rehabilitation services remotely. By design, telerehabilitation offers a practical alternative for stroke patients to overcome barriers such as limited access to specialized healthcare professionals, geographic constraints, and financial limitations [26]. Telerehabilitation provides the opportunity to remotely assist patients while they exercise at home through remote monitoring, addressing the problems of limited medical resources and staff, difficult transportation, or living a long distance from rehabilitation centers [26]. While telerehabilitation has demonstrated positive effects, is cost-effective, and increases access to rehabilitation, the primary barrier to its implementation is insufficient internet access [27]. Additionally, other obstacles, such as a lack of knowledge and familiarity with telerehabilitation, poor communication between professionals and stroke survivors, and technology-related issues, can also hinder the use of this strategy [28].

11.4.4 Self-Management Model

The self-management model in stroke rehabilitation is an emerging strategy for engaging stroke survivors in their own care [22]. This model aims to empower stroke survivors by encouraging them to take an active role in their recovery and daily lives. The self-management approach involves equipping patients with skills and strategies to address the spectrum of challenges after stroke [29–31]. The common challenges faced include depression, functional and mobility disability, reduction in life roles, and a lack of social support [29]. However, rehabilitation for stroke survivors is often targeted at improving physical and cognitive function, with less attention given to the psychosocial consequences of stroke. [30] To address these challenges, it's essential to implement rehabilitation strategies that help stroke survivors navigate the physical, emotional, and social difficulties they face post-stroke [30, 31]. These strategies should enable them to manage their health, lead fulfilling lives, and reach their maximum potential.

11.5 Real-World Implementation and Innovation

11.5.1 Brazil: Implementing Stroke Rehabilitation Care Network

Brazil is located in South America, with approximately 203 million inhabitants, making it the fifth-largest country in the world by population. The country is politically and administratively divided into 27 federative units. For political and operational purposes, the federative units are grouped into five regions (North, Northeast, Southeast, South, and Central-West), each with distinct geographical and cultural characteristics as well as significant socioeconomic disparities [32]. In 2022, the Northeast and North regions had the highest social vulnerability indexes (SVI) in the country, at 0.273 and 0.241, respectively [33]. The Brazilian healthcare system is universal, with approximately 70% of the population relying on it [34]. It is organized in a tripartite structure, with federal, state, and municipal governments jointly administering health care, including rehabilitation services.

In-hospital stroke rehabilitation is well established in Brazil due in part to the publication of ordinance 665 in 2012 [35]. This ordinance was responsible for establishing criteria for stroke centers. In type II stroke centers, it is mandatory to have daily support from a physiotherapist and speech therapist, while in comprehensive stroke centers, a multidisciplinary team is required [35]. The expansion of stroke centers across the country, since 2012, has increased access to reperfusion therapy but also to early in-hospital rehabilitation [36, 37].

However, after hospital discharge, the continuum of rehabilitation remains a significant challenge. A study conducted in a capital city in the Southeast region revealed that 66% of stroke patients received physiotherapy, 42% speech therapy, 17% occupational therapy, and none had access to psychological or social assistance services 1 month after hospital discharge [38]. The main barriers to accessing stroke rehabilitation were limited financial resources for health care, long wait times for scheduling, and scheduling process complexities [39]. Given socioeconomic disparities and the unequal distribution of rehabilitation professionals across Brazilian regions, the national reality could be even more concerning.

In light of these challenges, implementing effective strategies to organize the entire care network is essential. The city of Ribeirão Preto, located in the state of São Paulo in Southeastern Brazil, has a Stroke Unit since 2016 with a multidisciplinary team. Upon hospital discharge, all patients are referred to a physiatrist for an initial assessment and classification to determine the most appropriate type of rehabilitation for each individual. The city has a rehabilitation network that includes a Rehabilitation Hospital and outpatient services, with established criteria for patient admission and discharge (information collected through personal communication).

Investment in a care network organization has also yielded positive results in Joinville, located in the south of Brazil. Joinville has a monitoring program for all stroke cases in the city, called Joinvasc, which helps identify gaps and guide actions. The city also has a patient association and a comprehensive stroke unit with a dedicated multidisciplinary team. Stroke mortality decreased from 26.2% in 1995 to 10.6% in 2018–2019, and the proportion of severe stroke cases with improved functional status increased from 33.6% in 2010–2011 to 62.5% in 2018–2019, representing an 86% increase [40]. Despite these positive examples, it is important to emphasize that these cases do not necessarily reflect a national reality.

11.5.2 Malaysia: "Stroke Rehabilitation Continuum of Care" Initiative

Malaysia has launched an initiative to enhance stroke rehabilitation services by developing the "Stroke Rehabilitation Continuum-of-Care," or locally known as the *Program Kontinum Rehabilitasi Strok* (PKRS) [41]. This initiative aims to improve the quality of care and accessibility to rehabilitation for stroke survivors nationwide by strengthening stroke rehabilitation services in identified hospitals, including in nonurban localities. Spearheaded by the Ministry of Health Malaysia, this initiative is part of the stroke-care-related strategies outlined in the Medical Programme Strategic Planning for 2021–2025. The "Stroke Rehabilitation Continuum of Care" concept encompasses the transition of rehabilitation from the acute stroke phase and later into the community. This is in line with the World Health Organization (WHO) Rehabilitation Health Framework, which provides a common structure and organization of rehabilitation within healthcare [42]. The PKRS program, which is implemented in phases, focuses on providing comprehensive rehabilitation services, ensuring that patients receive timely and effective rehabilitation care. Efforts are in place to determine the implementation status and overall impact on the stroke patients.

11.5.3 India: "Care for Stroke" Innovation

An interesting example to showcase the utilization of the stroke rehabilitation models is the "Care for Stroke" innovation, which is a systematic effort to improve access to stroke rehabilitation information for the management of physical disabilities in India. "Care for stroke" was an intervention systematically coproduced with a range of stakeholders, including stroke survivors and families, to manage post-stroke disabilities in India [12]. This intervention was educational in nature, informed by a comprehensive needs assessment in the community. This was an amalgamation of three key strategies, namely CBR, self or supported management, and a technology-driven approach [43]. The intervention was web-based and enabled anyone with a smartphone to gain access to this intervention, which is organized in different sections, such as information about stroke, activities of daily living (ADL), home-based exercises, functional skills, and locally available assistive technology. The intervention consists of a range of 2–3 min videos related to sections with people affected by stroke, demonstrating actions related to these sections [44].

The "Care for Stroke" intervention was evaluated first for its feasibility and acceptability in an Indian context, and the results demonstrated that this intervention was highly acceptable and feasible to be implemented in India [45]. In recent years, this innovation was also evaluated for its clinical and cost-effectiveness through a randomized controlled trial. Results of this trial demonstrated that "Care for Stroke" intervention results in a protective effect of 40% when compared to any standard intervention [46]. Given the success of the intervention, it would be envisaged that this kind of intervention is a game changer to promote access to stroke services in LRS. However, the process evaluation, which is currently in progress, has informed that the stroke survivors, families, and care providers consider that this intervention may not be scalable because of the lack of organized pathways and systems for stroke care in the communities. Hence, raising an important implication for strengthening the systems for rehabilitation and care within the key agendas for health and social care in LRS. This intervention is currently translated for caregivers in Malaysia and is being evaluated for its scalability [47].

11.6 Building Collaborations and Peer Support Across Borders: (The South East Asia: The CMCL-WHO Collaboration, Tanzanian Stroke Project)

The Department of Neurology at the Christian Medical College and Hospital Ludhiana collaborates with the World Health Organization (WHO) as the globe's first-ever collaborating center for stroke care (treatment, prevention, and rehabilitation). Through this program, a team of stroke experts builds capacity among the relevant professionals in the WHO's South East Asian region (which includes 11 countries) in addition to advising leaders for inclusion of stroke care in national policies and practice guidelines, while also promoting systems of care and stroke-related research [3]. Such models of collaborations and peer support have shown improved outcomes both at the institutional and national levels. A similar program, but a small-scale national program in Tanzania, used a three-tier implementation approach. It aimed at improving the implementation of evidence-based stroke care in their country by developing national stroke registries, training healthcare workers, and establishing comprehensive stroke units. Such multilevel engagement with various stakeholders, introducing quality standards and monitoring of the same periodically, with efforts to sensitize healthcare workers on stroke rehabilitation, is the need in most LMICs [48].

11.7 Chapter Summary

Although stroke rehabilitation and treatment strategies have significantly advanced in recent years, the implementation and provision of stroke rehabilitation in resource-limited settings are challenging. Implementing effective stroke rehabilitation models in low-resource settings requires addressing numerous challenges, and it is imperative to first start at the ground level. This involves engaging with the local communities and educating them about the benefits of rehabilitation, which helps to dispel cultural misconceptions and encourage acceptance of these services. Public health campaigns and community involvement are equally essential to raise awareness and build local support. Other specific strategies include training the healthcare providers to provide stroke-related quality care and leveraging technology like telerehabilitation to overcome geographic and accessibility barriers. Emphasis shall also be made to advocate for an increase in funding and supportive policies to assist managers in organizing rehabilitation care networks. This effort will help to ensure financial sustainability to expand rehabilitation services and develop effective nationwide stroke care networks. Overcoming these barriers to improve access and increase the quality of stroke rehabilitation will ultimately depend on collaborative strategies that involve policymakers, stakeholders, healthcare professionals, professional societies, and civil society [12]. By employing these approaches, we can create a more inclusive and effective stroke rehabilitation environment, ultimately improving outcomes for stroke survivors even in the face of limited resources.

Stroke rehabilitation has progressed significantly, yet its implementation in resource-limited settings remains challenging. Effective models must begin at the grassroots level, engaging communities and educating them about the benefits of rehabilitation to dispel cultural misconceptions and foster acceptance. Public health campaigns and active community involvement are vital to raise awareness and build local support. Key strategies include training healthcare providers to deliver quality stroke care and using technologies like telerehabilitation to overcome geographic and access barriers. Advocacy for increased funding and supportive policies is essential to help health managers organize sustainable rehabilitation networks. These efforts aim to expand services and build nationwide systems that ensure continuity of care. Ultimately, improving access and quality in stroke rehabilitation depends on collaborative approaches involving policymakers, healthcare professionals, civil society, and professional organizations. By aligning efforts across sectors, it is possible to create a more inclusive, responsive, and effective rehabilitation environment that improves outcomes for stroke survivors, even in the face of limited resources.

What Does The Evidence Say

- Telehealth may be used as an alternative or complementary mode of rehabilitation delivery, particularly where access to specialist services is limited, provided interventions have proven clinical benefit.
- Early supported discharge should be offered to stroke survivors with mild to moderate disability where coordinated home-based services exist.
- Home-based rehabilitation is the preferred community model; if unavailable, center-based rehabilitation should be provided.
- Individualized, comprehensive discharge plans should be developed collaboratively with the stroke survivor and their caregiver before discharge.
- Stroke survivors and their families/carers should receive personalized information in appropriate formats, delivered at key stages of recovery.
- Ongoing engagement should include tailored materials, follow-up, clarification, and reinforcement.
- Stroke survivor-led self-management interventions should be introduced within the first 4 months post-discharge (e.g., "take charge after stroke").
- Cognitively able individuals and carers should be informed about and supported to access generic or stroke-specific self-management programs, with collaborative care plans where appropriate.
- Information about local stroke support groups and peer support resources should be provided both before discharge and during community reintegration.

(continued)

- Caregivers should receive continuous, tailored information and support across all stages of recovery, including details about the stroke, care team roles, investigations, treatment plans, discharge, and community services.
- Support may be delivered in person, via telephone, or through digital platforms.
- Carers should be offered ongoing psychological and social support, including strategies for coping, stress management, and problem-solving.
- Support should also focus on maintaining healthy relationships within the stroke survivor's support network.

References

1. Owolabi MO, Thrift AG, Martins S, Johnson W, Pandian J, Abd-Allah F, et al. The state of stroke services across the globe: report of world stroke organization-World Health Organization surveys. Int J Stroke. 2021;16(8):889–901.
2. Gonzalez-Aquines A, Rosales J, De Souza AC, Corredor-Quintero A, Barboza MA, Navia-Gonzalez V, et al. Availability and barriers to access post-stroke rehabilitation in Latin America. J Stroke Cerebrovasc Dis. 2024;33(10). Available from: https://pubmed.ncbi.nlm.nih.gov/39111374.
3. Sebastian IA, Gandhi DBC, Sylaja PN, Paudel R, Kalkonde YV, Yangchen Y, et al. Stroke systems of care in South-East Asia region (SEAR): commonalities and diversities. Lancet. 2023:17. Available from: https://pubmed.ncbi.nlm.nih.gov/37849930.
4. Roushdy T, Aref H, Kesraoui S, Temgoua M, Nono KP, Gebrewold MA, et al. Stroke services in Africa: what is there and what is needed. Int J Stroke. 2022;17(9):972–82. Available from: https://pubmed.ncbi.nlm.nih.gov/35034522.
5. de Fonoaudiologia CF. Quantitativo de Fonoaudiólogos no Brasil. 2023. http://fonoaudiologia.org.br/fonoaudiologos/quantitativo-de-fonoaudiologos-no-brasil-por-conselho-regional/.
6. Mlambo T, Pfavai Y, Chimusoro F, Kaseke F. Stroke rehabilitation and occupational therapy in low resource settings. In: Cerebrovascular diseases - elucidating key principles. London: IntechOpen; 2022. https://doi.org/10.5772/intechopen.101212.
7. Juckett L, Owolabi M, Kosir A, Ifejika N. Implementation science to advance health equity in stroke rehabilitation. J Am Heart Assoc. 2024;13:e031311. https://doi.org/10.1161/JAHA.123.031311.
8. Silva GS, Rocha E. Developing systems of care for stroke in resource-limited settings. Semin Neurol. 2024;44(2):119–29. https://doi.org/10.1055/s-0044-1782617. Epub 2024 Mar 21. PMID: 38513704.
9. Gururaj S, Bird ML, Borschmann K, Eng JJ, Watkins CL, Walker MF, et al. Evidence-based stroke rehabilitation: do priorities for practice change and feasibility of implementation vary across high income, upper and lower-middle income countries? Disabil Rehabil. 2022;44(17):4611–8. Available from: https://pubmed.ncbi.nlm.nih.gov/33849357.
10. Feigin VL, Owolabi MO, Owolabi MO, Owolabi MO, Feigin VL, Abd-Allah F, et al. Pragmatic solutions to reduce the global burden of stroke: a world stroke organization-lancet neurology commission. Lancet Neurol. 2023;22(12):1160–206.
11. Cormican A, Hirani SP, McKeown E. Healthcare professionals' perceived barriers and facilitators of implementing clinical practice guidelines for stroke rehabilitation: a systematic review. Clin Rehabil. 2023;37(5):701–12.
12. Owolabi MO, Suwanwela NC, Yaria J. Barriers to implementation of evidence into clinical practice in low-resource settings. Nat Rev Neurol. 2022;18(8):451–2.

13. WHO. Community-based rehabilitation: CBR guidelines. Geneva: World Health Organization; 2010.
14. Iemmi V, Kumar KS, Blanchet K, Gibson L, Hartley S, Murthy GVS, Patel V, Weber J, Kuper H. Community-based rehabilitation for people with physical and mental disabilities in low- and middle-income countries. Cochrane Database Syst Rev. 2017;2017(3):CD010617. https://doi.org/10.1002/14651858.CD010617.pub2.
15. Robertson J, Emerson E, Hatton C, Yasamy M. Efficacy of community-based rehabilitation for children with or at significant risk of intellectual disabilities in low- and middle-income countries: a review. J Appl Res Intellect Disabil. 2012;25(2):143–54.
16. Iemmi V, Gibson L, Blanchet K, Suresh Kumar K, Rath S, Hartley S, Murthy GV, Patel V, Weber J, Kuper H. Community-based rehabilitation for people with disabilities in low- and middle-income countries: a systematic review. Campbell Syst Rev. 2015;15:177.
17. Kamalakannan S, Srivastava A, Chockalingam M, et al. Health systems for rehabilitation - critical appraisal of the National Programme for stroke in India. J Indian Feder Neurorehabil. 2024;1(1):18–26.
18. Kamalakannan S, Gudlavalleti Venkata M, Prost A, Natarajan S, Pant H, Chitalurri N, Goenka S, Kuper H. Rehabilitation needs of stroke survivors after discharge from hospital in India. Arch Phys Med Rehabil. 2016;97(9):1526–32. https://doi.org/10.1016/j.apmr.2016.02.008.
19. Sebastian IA, Demers M, Yeghiazaryan N, et al. Establishing organized stroke care in low- and middle-income countries: from training of non-specialist to implementation. J Stroke Med. 2023;6(1):46–51. https://doi.org/10.1177/25166085231158425.
20. Pandian JD, Kalkonde Y, Sebastian IA, Felix C, Urimubenshi G, Bosch J. Stroke systems of care in low-income and middle-income countries: challenges and opportunities. Lancet. 2020;396(10260):1443–51. https://doi.org/10.1016/S0140-6736(20)31374-X.
21. Kovacevic P. Can the terms "low resource setting" and "low-income country" be used interchangeably in the context of intensive care medicine? Intensive Care Med. 2023;49:1274–5. https://doi.org/10.1007/s00134-023-07203-8.
22. Ekechukwu END, Olowoyo P, Nwankwo KO, Olaleye OA, Ogbodo VE, Hamzat TK, Owolabi MO. Pragmatic solutions for stroke recovery and improved quality of life in low- and middle-income countries-a systematic review. Front Neurol. 2020;11:337. https://doi.org/10.3389/fneur.2020.00337.
23. Joshi R, Alim M, Kengne AP, Jan S, Maulik PK, Peiris D, et al. Task shifting for non-communicable disease management in low- and middle-income countries-a systematic review. PLoS One. 2014;9:e103754. https://doi.org/10.1371/journal.pone.0103754.
24. Lindley RI, Anderson CS, Billot L, Forster A, Hackett ML, Harvey LA, et al. Family-led rehabilitation after stroke in India (ATTEND): a randomised controlled trial. Lancet. 2017;390:588–99. https://doi.org/10.1016/S0140-6736(17)31447-2.
25. Forster A, Dickerson J, Young J, Patel A, Kalra L, Nixon J, et al. A structured training programme for caregivers of inpatients after stroke (TRACS): a cluster randomised controlled trial and cost-effectiveness analysis. Lancet. 2013;382:2069–76. https://doi.org/10.1016/S0140-6736(13)61603-7.
26. Ciortea VM, Motoaşcă I, Ungur RA, Borda IM, Ciubean AD, Irsay L. Telerehabilitation—a viable option for the recovery of post-stroke patients. Appl Sci. 2021;11(21):10116. https://doi.org/10.3390/app112110116.
27. English C, Ceravolo MG, Dorsch S, Drummond A, Gandhi DBC, Halliday Green J, et al. Telehealth for rehabilitation and recovery after stroke: state of the evidence and future directions. Int J Stroke. 2022;17(5):487–93. Available from: https://pubmed.ncbi.nlm.nih.gov/34983266.
28. Nizeyimana E, Joseph C, Plastow N, Dawood G, Louw QA. A scoping review of feasibility, cost, access to rehabilitation services and implementation of telerehabilitation: implications for low- and middle-income countries. Digit Health. 2022. Available from: https://pubmed.ncbi.nlm.nih.gov/36249480.

29. Parke HL, Epiphaniou E, Pearce G, Taylor SJ, Sheikh A, Griffiths CJ, et al. Self-management support interventions for stroke survivors: a systematic meta-review. PLoS One. 2015;10:e131448. https://doi.org/10.1371/journal.pone.0131448.
30. Jones F, Riazi A, Norris M. Self-management after stroke: time for some more questions? Disab Rehabil. 2013;35:257–64. https://doi.org/10.3109/09638288.2012.691938.
31. Lennon S, McKenna S, Jones F. Self-management programmes for people post stroke: a systematic review. Clin Rehabil. 2013;27:867–78. https://doi.org/10.1177/0269215513481045.
32. IBGE. Censo Demográfico | IBGE [Internet]. 2022 [cited 2024 Oct 29]. Available from: https://censo2022.ibge.gov.br/panorama/.
33. Funari AP, Dias TTR, Teixeira LCM, Costa MA. Atualização do Índice de Vulnerabilidade Social (IVS) a partir dos dados da PNAD contínua 2022: breve comentário metodológico e resultados gerais. Boletim Reg Urbano Ambiental. 2024;31:77–84.
34. IBGE. Pesquisa nacional de saúde: 2019: percepção do estado de saúde, estilos de vida, doenças crônicas e saúde bucal: Brasil e grandes regiões. Rio de Janeiro; 2020.
35. Ministério da Saúde. Portaria no 665 de abril de 2012. 2012.
36. Martins SCO, Pontes-Neto OM, Alves CV, de Freitas GR, Filho JO, Tosta ED, et al. Past, present, and future of stroke in middle-income countries: the Brazilian experience. Int J Stroke. 2013;8:106–11.
37. Ouriques Martins SC, Sacks C, Hacke W, Brainin M, de Assis FF, Marques Pontes-Neto O, et al. Priorities to reduce the burden of stroke in Latin American countries. Lancet Neurol. 2019;18:674–83.
38. Magalhães JP, Faria-Fortini I, Dutra TM, Sant'Anna R, Soares C, Teixeira-Salmela LF, et al. Access to rehabilitation professionals by individuals with stroke one month after hospital discharge from a stroke unit in Brazil is insufficient regardless of the pandemic. J Stroke Cerebrovasc Dis. 2023;32(8):107186.
39. Soares CLA, Magalhães JP, Faria-Fortini I, Batista LR, Andrea Oliveira Lima L, Faria CD. Barriers and facilitators to access post-stroke rehabilitation services in the first six months of recovery in Brazil. Disabil Rehabil. 2024;46(23):5601–7.
40. Diegoli H, Magalhães P, Makdisse M, Cabral Moro CH, França P, Longo A. Joinvasc: organizing the full cycle of stroke care with universal coverage. NEJM Catal Innov Care Deliv. 2022;4(1):283.
41. Modul program kontinum rehabilitasi stroke (PKRS) 1st ed. 2023.
42. Rehabilitation in health systems. Guide for action. Geneva: World Health Organization; 2019.
43. Sureshkumar K, Murthy GV, Kinra S, Goenka S, Kuper H. Development and evaluation of a smartphone-enabled, caregiver-supported educational intervention for management of physical disabilities following stroke in India: protocol for a formative research study. BMJ Innov. 2015;1(3):117–26. https://doi.org/10.1136/bmjinnov-2015-000042.
44. Sureshkumar K, Murthy GV, Munuswamy S, Goenka S, Kuper H. 'Care for Stroke', a web-based, smartphone-enabled educational intervention for management of physical disabilities following stroke: feasibility in the Indian context. BMJ Innov. 2015;1(3):127–36. https://doi.org/10.1136/bmjinnov-2015-000056.
45. Sureshkumar K, Murthy G, Natarajan S, Naveen C, Goenka S, Kuper H. Evaluation of the feasibility and acceptability of the 'care for stroke' intervention in India, a smartphone-enabled, carer-supported, educational intervention for management of disability following stroke. BMJ Open. 2016;6(2):e009243. https://doi.org/10.1136/bmjopen-2015-009243.
46. Sureshkumar K, Murthy GVS, Kuper H. Protocol for a randomised controlled trial to evaluate the effectiveness of the 'Care for Stroke' intervention in India: a smartphone-enabled, carer-supported, educational intervention for management of disabilities following stroke. BMJ Open. 2018;8(5):e020098. https://doi.org/10.1136/bmjopen-2017-020098.

47. Sidek NN, Tengku Ismail TA, Kamalakannan S, Chen XW, Romli MH, Mat Said MZ, Papachristou Nadal I, Ibrahim KA, Musa KI. A mixed-methods study on the implementation of a mobile health application (mHealth app) for stroke caregivers in Malaysia: healthcare providers' perspective. Front Neurol. 2023;14:1222260. https://doi.org/10.3389/fneur.2023.1222260.
48. Matuja SS, Tunkl C, Roushdy T, Li L, Ouyang M, Saleh Velez FG, Gebrewold M, et al. Implementing acute stroke services in sub-saharan africa: steps, progress, and perspectives from the Tanzania stroke project. Cerebrovasc Dis Extra. 2025;15(1):143–53. https://doi.org/10.1159/000545954.

12 Operating Guides for In-Hospital and Community-Based Stroke Rehabilitation in LMICS

Dorcas B. C. Gandhi

Abstract

Stroke is a leading cause of long-term disability worldwide, with a disproportionate burden in low- and middle-income countries (LMICs). Effective rehabilitation is essential for improving functional outcomes, reducing dependency, and enabling community reintegration for stroke survivors. This chapter outlines practical operating guides and standard operating procedures (SOPs) for implementing stroke rehabilitation across hospital and community settings in LMICs, with particular relevance to India.

The chapter describes structured approaches to in-hospital stroke rehabilitation, beginning with early referral, multidisciplinary team (MDT) coordination, and rehabilitation assessment within 24–48 h of medical stabilization. Emphasis is placed on ICF-based multidimensional assessment, early task-specific therapy, goal-oriented rehabilitation planning, and structured caregiver training. In settings with a limited workforce, task-sharing and family-led rehabilitation models are highlighted as pragmatic strategies to extend rehabilitation coverage. Standardized discharge planning, including home exercise programs, patient education, and referral to community resources, is recommended to ensure continuity of care.

This chapter also presents community- and home-based rehabilitation models, which are critical in LMIC contexts where institutional rehabilitation capacity is limited. Home assessments, individualized rehabilitation plans, and integration with primary care and community-based rehabilitation (CBR)

D. B. C. Gandhi (✉)
Department of Neurology, Christian Medical College & Hospital Ludhiana, Ludhiana, Punjab, India

Department of Physiotherapy, Manipal Academy of Higher Education, Manipal, Karnataka, India
e-mail: dorcas.gandhi@cmcludhiana.in

D. B. C. Gandhi, J. D Pandian (eds.), *Stroke Rehabilitation in Low-Resource Settings*, https://doi.org/10.1007/978-981-92-0403-8_12

systems are emphasized. Training community health workers and linking stroke rehabilitation with non-communicable disease services can enhance long-term care and prevention. Structured follow-up schedules and simple quality indicators are proposed to monitor outcomes and strengthen rehabilitation systems.

Overall, the chapter provides practical, scalable guidance for integrating stroke rehabilitation into health systems across hospital and community levels. By combining evidence-based practices with context-appropriate adaptations, these SOPs aim to support the expansion of accessible, effective stroke rehabilitation services in LMICs.

Keywords

Stroke rehabilitation • Low- and middle-income countries (LMICs) • Standard operating procedures • Multidisciplinary team • Family-led rehabilitation • Community-based rehabilitation (CBR) • Home-based rehabilitation • Stroke care pathways • Rehabilitation 2030 initiative

WHO's rehabilitation 2030 initiative requires rehabilitation, including stroke rehabilitation, to be implemented as an essential health service across all levels of the health system globally. This means rehabilitation should not be an option, but an integral part of healthcare delivery. This chapter outlines the various evidence-based tools and standard operating procedures (SOPs) for stroke rehabilitation implementation:

1. In-hospital stroke rehabilitation, from admission to discharge.
2. Community-based and home-based stroke rehabilitation, including early supported discharge and family-led models.

Key principles in setting evidence-based stroke rehabilitation services in low- and middle-income countries (LMICs):

- Stroke unit care with early, coordinated rehabilitation reduces death and dependency [1].
- Multidisciplinary team (MDT) working, structured goal setting, and intensive, task-specific practice improve functional outcomes and patient satisfaction [2].
- Early supported discharge (ESD) with coordinated home-based rehabilitation can reduce length of stay and improve independence for selected patients, when multidisciplinary community services exist [3].
- In LMICs, evidence supports exercise-based and cognitive training interventions, delivered in low-resource settings, as effective for improving post-stroke outcomes [4].
- Given workforce shortages, task-sharing and family-led rehabilitation are realistic and evidence-supported adaptations; the Indian ATTEND trial is a major example [5].

- Systematic reviews of stroke rehab guidelines in LMICs highlight gaps in scope, quality, and implementation but broadly endorse the same core components as high-income guidelines, adapted to context [6].

12.1 SOPs for in-Hospital Stroke Rehabilitation

12.1.1 Service Organization: Stroke-Capable Units and MDTs

Even in the absence of fully functional stroke units (SUs), hospitals should implement stroke-capable wards with structured processes and protocols. The minimum standards for such units in a low-resource setting should include:

1. Dedicated stroke care area (beds in one ward or bay where possible).
2. A multidisciplinary team including, but not limited to, a neurologist/physician, stroke nurse, stroke-trained physiotherapist, and medical social worker.

   Evidence has shown that, with workforce limitations in LMICs, it is ideal for stroke nurses to take over speech/language and swallow rehabilitation in collaboration with the caregivers, while occupational therapy demands and coordination with the social worker for smooth transitioning to the community can fall under the purview of the stroke-trained physiotherapist. Such modified teams should ideally meet once a week to review all stroke inpatients, set goals, and plan discharge.

   A named stroke rehab coordinator (often PT or rehab nurse) responsible for tracking each patient's rehab pathway is ideal; however, in low-resources settings, either the nurse or physician-assistants can take over this role.
3. Referral SOP

   It is essential that an automatic rehab referral for all confirmed stroke admissions (ischemic or haemorrhagic) is practised regardless of initial severity or need for rehabilitation. This could be facilitated by including a stroke rehab referral checkbox in admission and transfer forms, enabling nurse-initiated referrals to rehab if the form is incomplete within 48 h of admission.
4. Time target:

   An ideal SU should aim to complete the rehab assessment within 24–48 h of admission/medical stabilization.

12.1.2 Assessment and Therapy: Function, Risk, and Context

An ICF-based, multidimensional assessment lays the foundation for realistic goal setting with a target of initial assessment within 24–48 h of admission/stability. Such a process should include recommended sensorimotor outcome measures as described by Kwakkel et al. in 2017 [7], Swallowing screening and communication

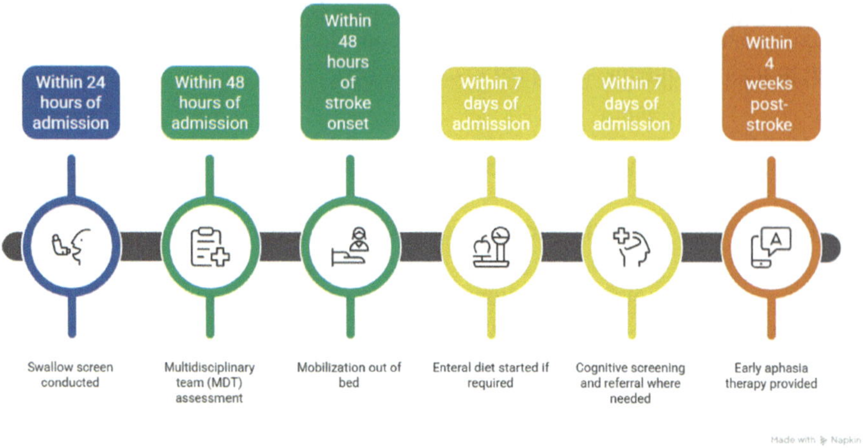

Fig. 12.1 Time-critical stroke rehabilitation milestones

assessment for all with suspected bulbar or language involvement (refer to Chap. 6), and cognitive status screening (Chap. 5) at least.

The stroke coordinator or social worker should document the risk profile which should include, falls risk (alternately docne by the PT), risk of aspiration, pressure injuries, DVT, depression/anxiety screening, social and environmental context (type of housing, stairs, distance from hospital), availability of caregiver(s), their health and literacy, financial constraints, and insurance/public coverage.

All assessment findings should ideally be documented in a standardized stroke rehab assessment form, ideally integrated with the medical record.

Stroke rehabilitation should prioritize appropriate therapy content and sufficient intensity, ensuring early initiation and high-frequency, task-based repetitions. The following image can be used as a guide in in-hospital acute settings for planning and initiating various domains of rehabilitation (Fig. 12.1)

12.1.3 Goal Setting and Treatment Planning

Guidelines and observational evidence emphasize person-centred, goal-oriented rehab to improve engagement and outcomes. Goal-setting within 3–5 days of admission with all relevant stakeholders (patients, caregiver(s), MDT representatives) is important. A structured set of SMART goals should be explored, prioritizing patient preferences while balancing therapy outcomes and baseline findings. Clear documentation and communication to all relevant members is essential. Where feasible, assigning a responsible team member and an expected time frame for each goal would be beneficial.

12.1.4 Caregiver Training and Family-Led Rehabilitation

The ATTEND trial from India tested a structured family-led stroke rehabilitation programme, where family caregivers were trained to deliver evidence-based exercises and functional activities under professional guidance. Although ATTEND did not show a significant reduction in death or dependency at 6 months compared with usual care, it demonstrated that family-led rehabilitation is feasible, safe, and acceptable, and can partially compensate for limited professionals. Studies are currently exploring modified models of caregiver-led rehabilitation delivery with improved and prolonged remote supervision by qualified professionals. Such models, although not proven completely by research, are the low-hanging fruit that an already stretched health system can rely on for improved reach.

While implementing such a rehab-delivery model, the following aspects should be considered:

- Identification of a primary caregiver (or two) early.
- Inclusion of caregivers in rehab sessions within the first week of admission.
- Provision of structured training sessions covering:
  - Safe transfers and mobility assistance
  - Positioning, exercise routines, and the use of aids
  - Communication strategies and cognitive engagement
  - Recognition of red-flag symptoms (worsening weakness, chest pain, new confusion)
  - Use simple printed or pictorial materials in the local language.
- Observe and correct caregiver performance before discharge.
- Where feasible, leverage patient-recorded exercise videos to objectively compare functional progress across successive months.

12.1.5 Discharge Planning and Early Supported Discharge

It is crucial for stroke survivors to leave the hospital with sufficient guidance in continued rehabilitation, secondary prevention, and follow-up plans while educating them in traversing functional needs within their home and community set-up. Even without a full discharge team, hospitals can standardize discharge processes by:

- Reassessing function at discharge (functional scales/mRS, communication, swallowing).
- Developing a home programme (written or audio-visual):
  - Exercises (frequency, repetitions, position) with pictures
  - Mobility and safety precautions
  - Swallowing and diet instructions
  - Daily activity plan (practice walking, self-care, and communication)
- Providing a structured discharge summary:

- Stroke type, treatments, residual deficits
- Functional status and rehab goals
- Follow-up appointments and referrals
- Link to community resources
- Identifying local rehab services, community-based rehabilitation (CBR) programmes, government disability services, and self-help groups where they exist.
- Providing contact details and formal referral notes.
- Planning follow-up:
  - Schedule review (e.g. 4–6 weeks and 3 months)
  - Where feasible, arrange phone or tele-rehab check-ins

12.2 SOPs for Community- and Home-Based Stroke Rehabilitation

With limited institution-based capacity, LMICs push rehabilitation beyond the hospital, using home- and community-based models that integrate with primary care and community health workers. For more details on service delivery models, please refer to Chap. 11.

12.2.1 Community Assessment and Individual Rehab Plans

Assessments in the home/community should focus on practical function and environmental barriers.

Home assessment should ideally be conducted by the social worker trained in stroke protocols, visit within 1–2 weeks of discharge. Such an evaluation should include:

- Mobility in the actual environment (getting in/out of bed, toilet, navigating stairs/uneven ground)
- ADLs (feeding, bathing, dressing, toileting)
- Communication in real-life interactions
- Participation: ability to attend temple/church/mosque, local meetings, or community events
- Identify environmental barriers
- Narrow doors, steps, the absence of handrails, poor lighting, unsafe bathrooms
- Social/attitudinal barriers (stigma, overprotection)

A thorough review of adherence to the home exercise programme and the correctness of technique is essential. This can be achieved by using standardized exercise performance videos, both live and pre-recorded, maintained in an institutional repository. Such reference standards enable consistent assessment, reinforce correct technique, and improve patient compliance.

Based on the home assessment, an individual rehabilitation plan can be updated with the guidance of qualified rehabilitation experts. This plan should include two to three prioritized goals (e.g. "walk safely to toilet", "feed self", "attend weekly worship"), specific actions for patient and caregiver, community worker, and specialist therapists (during periodic visits or tele-consults).

12.2.2 Linkages with Primary Care and CBR Structures

In India and similar settings, it is logical to embed stroke rehabilitation into existing CBR or community health platforms by

- Training ASHAs, primary care doctors, and CBR workers using short, competency-based modules on stroke identification, basic rehab, and prevention of complications.
- Integrating stroke rehab review into non-communicable disease (NCD) clinics, as stroke survivors often have hypertension, diabetes, and heart disease that need long-term management.
- Aligning with disability and social protection schemes, ensuring survivors have access to disability certificates, pensions, and assistive device programmes.

12.2.3 Follow-Up, Monitoring, and Escalation

At the bare minimum, follow-ups should be scheduled at 15 days, 1 month, 3 months, 6 months, and 12 months after stroke. At each of these time points, tracking the following is crucial:

- Functional status.
- Participation.
- New medical complications (falls, seizures, pain, contractures, pressure sores).
- Escalate to hospital or specialist review if there is sudden deterioration, new symptoms, or plateau/regression despite good adherence.

Data from such follow-up can feed into local and national information systems, closing the current data gap on post-stroke disability in LMICs.

12.2.4 Quality Assurance

At an institutional and national level, quality indicators can serve to improve the standards of care. Core quality indicators that even low-resource services can track are:

- Proportion of stroke inpatients receiving rehab assessment within 48 h of stabilization.
- Average number and duration of therapy sessions (PT/OT/SLP) per inpatient per week.
- Proportion discharged with a written home programme and documented caregiver training.
- Proportion with at least one community follow-up contact within 1 month of discharge.
- Change in simple function scores from admission to 3 months.

These basic metrics align with the broader Rehabilitation 2030 agenda of data-driven strengthening of rehab within health systems and are crucial to represent the population from LMICs in big data analysis when planned/conducted. Keep a look-out for the quality metrics of the World Stroke Organization in implementating stroke rehabilitaiton at minimal, essential and advance levels of stroke units.

References

1. Bernhardt J, Urimubenshi G, Gandhi DB, Eng JJ. Stroke rehabilitation in low-income and middle-income countries: a call to action. Lancet. 2020;396(10260):1452–62.
2. Royal College of Physicians, Intercollegiate Stroke Working Party. National clinical guideline for stroke. London: Royal College of Physicians; 2023. Available from: https://www.strokeguideline.org/chapter/rehabilitation-and-recovery-principles-of-rehabilitation/?id=228. Accessed 19 January 2026.
3. National Institute for Health and Care Excellence (NICE). Stroke rehabilitation in adults. London: NICE. Available from: https://www.nice.org.uk/guidance/ng236.
4. Dee M, Lennon O, O'Sullivan C. A systematic review of physical rehabilitation interventions for stroke in low and lower-middle income countries. Disability Rehabil. 2020;42(4):473–501.
5. Lindley RI, Anderson CS, Billot L, Forster A, Hackett ML, Harvey LA, Jan S, Li Q, Liu H, Langhorne P, Maulik PK. Family-led rehabilitation after stroke in India (ATTEND): a randomised controlled trial. Lancet. 2017;390(10094):588–99.
6. Gandhi DB, Kamalakannan S, Urimubenshi G, Sebastian IA, Montanaro VV, Chawla NS, D'souza JV, Ngeh E, Mahmood A, Demers M, Hombali A. Stroke rehabilitation clinical practice guidelines in low-and middle-income countries: a systematic review of quality and unique features. Cerebrovasc Dis. 2025;54(4):429–40.
7. Kwakkel G, Lannin NA, Borschmann K, English C, Ali M, Churilov L, Saposnik G, Winstein C, Van Wegen EE, Wolf SL, Krakauer JW. Standardized measurement of sensorimotor recovery in stroke trials: consensus-based core recommendations from the stroke recovery and rehabilitation roundtable. Int J Stroke. 2017;12(5):451–61.

GPSR Compliance

The European Union's (EU) General Product Safety Regulation (GPSR) is a set of rules that requires consumer products to be safe and our obligations to ensure this.

If you have any concerns about our products, you can contact us on ProductSafety@springernature.com

In case Publisher is established outside the EU, the EU authorized representative is:

Springer Nature Customer Service Center GmbH
Europaplatz 3
69115 Heidelberg, Germany

Batch number: 10889560

Printed by Printforce, the Netherlands